E.M.C. Press Ltd.
4 Denby Drive Mansfield
Nottinghamshire NG18 4SS England
Phone 01623 624899 Mobile 07803 750295
Phone/Fax 01623 648575
e-mail wordtrak@yahoo.co.uk
Company Registration No. 5350293

This edition published by
E.M.C. Press Ltd. 2006

A copy of the British Library Cataloguing in
Publication Data is available from the British Library

ISBN 0-9554525-0-3
(978-0-9554525-0-5)

Printed and bound in the UK by
Wyndeham-Gait (Printers) Grimsby.

About the Author

Roy Bainton was born a long time ago in Hull, East Yorkshire. He moved to Mansfield in 1986. He has published six books covering a variety of subjects including popular music, biography and modern history. He has also written extensively for BBC Radio and numerous publications including *The New Statesman, Daily Mail, The Observer, The Independent, Saga Magazine* and many others, and has worked as TV scriptwriter for programmes on Sky's Life Channel.

SPECIAL THANKS:

We offer our sincere gratitude to Diana LeCore of the Society of Indexers for her diligent work in preparing the index for this book, and special thanks to Tony Corn of Wyndeham Gait in Grimsby, for his professionalism and attention to detail throughout the printing process.

The Alternative Mansfield Brew

The Story of a Great Brewery – Told by the men and women who made the beer.

By

R O Y B A I N T O N

E.M.C. Press Ltd. 4 Denby Drive
Mansfield Notts. NG18 4SS

Phone 01623 624899 Mobile 07803 750295 Phone/Fax 01623 648575
e-mail wordtrak@yahoo.co.uk
Company Registration No. 5350293

CONTENTS

Foreword
By
John Else, former Director of Personnel, Mansfield Brewery Plc.

There were four of us. John Bartle, Sam Sills, both of whom had worked in Mansfield Brewery's Wines and Spirits department, myself, and Graham Thorman, from the company's accounts department.
In 1995 this lucky quartet embarked on what was to be an annual series of visits to wine regions. The first was to Oporto. It was organised by IDV (International Distillers and Vintners). Our task was to sample Port, see its production and marketing. We visited the Port lodges, were treated and entertained like Kings, and ate superb food in restaurants. We were taken up to Quinta Rueda in the mountains and were shown around the vineyards and the winery. We drank 1994 Port, which looked, by its colour, bright purple, to be undrinkable. Yet it was delicious, and the '94 vintage has scored 10/10 ever since. The following year, again courtesy of IDV, we all went to Jerez, and sampled Croft Port and Gonzalez Byass sherry. We stayed in an apartment in Porto Santa Maria, and ate out at night. In 1997 IDV sent us willingly to France to stay in a gite in Burgundy. We would sample the wines of the region as part of our mission. One house, Prieur Brunet was particularly memorable. The owner and her husband showed us great hospitality in their dining room. He asked us if we had tasted various vintages and when we said we hadn't, he obliged. When we'd finished the meal there were 14 bottles on the table. We had the impression that Madame's patience had been overstretched, but it was a great visit.
Back at the gite, among ourselves, over copious bottles of excellent wine, the tales of life at the Brewery began to unfold. I recall Sam Sills saying, one night, 'You know, all this should be in a book'. From

then on, every time we reminisced, that phrase was repeated. During these trips the demarcation among us was simple: I did all the driving, Sam did the cooking, John Bartle did all the cleaning up and Graham, well, Graham read a book – he was 'on holiday'.
A couple of years later, Sam had a stroke and we didn't make a trip. But a decision was made; we had to turn the dream of this book into a reality. So, for Sam, John, Graham and all the employees of Mansfield Brewery, here's the dream – a reality at last.

I left school in 1959. I went into engineering. I got as far as ONC level at college studying engineering, but left it behind to join the entertainment industry. My entertainment partner was Stan Crosby from Mansfield Woodhouse. His father had the shop, café and hall at the corner of High Street in Woodhouse - in fact it was known as Crosby's Corner. I managed a rock band. As a group we started the Sutton In Ashfield Sutton Hop at St. Michael's Church Hall, which was extremely successful. Stan and I brought out the remainder of the group who were running the dances. This went on to forming an agency, selling groups all over the UK and into Germany. The worst loss was buying and then selling The Searchers. We lost £900 – we didn't get paid. It was a lot of money then. It was all great fun while it lasted but eventually I left the entertainment business and went back to college to specialise in Work Study, Materials Handling, Training and Personnel. I went on to work in Training for Mansfield Hosiery Mills and someone passed me an advert for a job with Mansfield Brewery.

In 1975, at the age of 35 I started as Training Manager with Mansfield Brewery on a split contract with Hardy's & Hansons Brewery. Unbeknown to me, in 1978 Mansfield Brewery were making a bid for Shipstone's Brewery. Because of this Mansfield said that they needed a full time Training Manager and asked me which company I would like to work for, Hardy's or Mansfield, although I didn't know about the bid for Shipstone's. I chose Mansfield Brewery. My preference for Mansfield Brewery was based on the go-ahead nature of Mansfield Brewery compared to that (at the time) of Hardy's and Hansons. I

thought that from a Training and Personnel point of view Mansfield would go forward much quicker. This was proven correct. However, the thing that didn't happen was the take-over of Shipstone's. As a knock-on effect, the Personnel Manager of Mandora, Mansfield Brewery's soft drinks company, lost his job, which was very unfortunate. My job then developed into Training and Personnel Manager. I dealt with everyone from the bottom up excluding the Senior Managers and Directors. The new Managing Director, John Hings, when he joined, split the company up into divisions, with boards of directors for each division. I eventually became a Divisional Director of the Trading Company. The Trading Company was brewing, packaging, free trade selling, warehousing and distribution. Other divisions were Mansfield Inns, which ran the pubs, Mandora Soft Drinks, and Mansfield Brewery Plc, the main board. At one point, I became responsible for Personnel in all the divisions, and then my title became Head of Personnel.
Whether or not it was to do with Mansfield Brewery producing beer and beer was all about pubs and entertainment, I don't know, but working for the Brewery was far from being just any job. It was something special. I could see myself working there until retirement – I wanted to, and so I did. My contract allowed me to retire at 60. I imagined retirement to be that time when my wife and I would be able to do all the things we'd planned. After seven years of retirement, none of it has happened. During almost 25 years with the company I have made many close friends and we're all in regular contact. It's referred to quite accurately as 'the Mansfield family'. We still participate in activities we used to participate in at the Brewery.

I only heard about the impending take-over just hours before the Directors were to make a presentation on the subject. Managing Director Bill McCosh took me into his confidence. I was shocked and stunned. Once the shock had subsided, I made a statement; I simply said 'They can't do that'. I asked Bill McCosh if, after he'd made his presentation to the senior managers, I could call a meeting with them to discuss what action we might take together, because as far as I was concerned the Trust Deed prevented the Trust from selling the shares. He said I could. Needless to say they were stunned, too. The move to sell was activated by the Baily Thomas Trust. Later in this book you will read the fascinating story of how this crucial trust which, among other things, was important in maintaining the company's independence, was set up. I suspect that the idea of selling Mansfield Brewery had its origins ten years earlier when there had been a major fall-out between the two trustee representatives on the board, Robin Chadburn, the company Chairman, and Jock Nangle, a non-executive director who was representative for the Trust. They fell out quite seriously over the direction and future stewardship of the company. I think that could well have been the seed which grew eventually into the actual sell-off. Both trustees could nominate their own successors. They both did, and nominated their sons. Robin Chadburn nominated Richard Chadburn and Jock Nangle his son Jocelyn. That apart, the breakdown between Robin and Jock at the time split the board. Robin Chadburn and Jock Nangle both retired and other board members left. I have no idea if my theory is true, but it is my belief that at the time of the disagreement the decision was taken that when the man who held 50% of the company, Frank Baily Thomas (whose sad story is included in this book) eventually died, then that would be the time to part company with the ownership of Mansfield Brewery. Frank Baily Thomas died in 1997 and by 1999 the company had gone, a period which included a six-month fight to try and prevent the sale. It seemed that on Frank's death, the plan could be activated. And it was. I think everyone was surprised that poor Frank lived for so long.

The main contributors on the committee set up to try and save the Brewery were myself, Paul Cooper, Richard Meadows, Tony Gilbert, Neil Tomlinson and Terry Johnson. After we got the fight going we realised we needed cash, so we set about fund raising. Everyone at the Brewery contributed to the fighting fund from the shop floor upwards. We raised £77,000 which came from employees and generous donations from suppliers. I think the biggest donation was £5,000 and the lowest £1. The court case went ahead with both sides stating their case. Listening to the judge summing up, Paul and I thought we'd done pretty well. We even thought we might win. Unfortunately, it then all turned on its head. We began to realise we'd probably lost. But that last day we still had hope. The legal bill and the report we commissioned came to £265,000. Fortunately we got the Provident Fund to pay most of this, except for £17,000. Everyone got their money back in proportion to what they'd donated, except for those donors we couldn't track down, so their money was given to charity.

That trip home after losing the court case was very depressing. It felt like a death in the family. It was approaching Christmas, the worst time for a family death. To give Bill McCosh his credit, the redundancy packages were second to none, but everyone would have willingly changed theirs for their job back. I was the first man out after the main board. I left on February 4th 2000. The other workers went in waves. The office staff went first, production in 2002. One good thing the trustees did work into the deal was that 75% of the workforce would be kept on for at least 2 years. The state-of-the-art brewing equipment in the brewhouse was transferred to Marston's in Burton, where it now operates.

The final decision which sealed the Brewery's fate was made by men in wigs. Not one of the people involved with the legal profession showed one bit of interest in the second clause in the trust document which stated categorically that Mansfield Brewery should be kept independent. Nobody was interested in that. As far as they were concerned it wasn't a legal issue. Despite its huge visible success down the decades, the difficulty with Mansfield Brewery was that its growth was restricted. The Trusts wouldn't let their shareholding be diluted. The Board were fenced in – yet within that fence, becoming the 7th largest brewery in the UK, they operated magnificently.

Mansfield Brewery has been the major, most enjoyable part of my working life and, with my continuing connections today with all the ex-employees it still is.

Compiling this book has been a labour of love. It takes me back to those happy nights in Porto and Burgundy with Sam, John and Graham when we decided that the dream of this book had to be realised. For my colleagues, for all the employees down the years, and for you, the readers, the story of Mansfield Brewery Plc wasn't a dream – it was a wonderful reality. I hope you enjoy reading about it as much as we enjoyed living it.

John Else,
Mansfield November 2006.

The Alternative Mansfield Brew

He that buys land buys many stones;
He that buys flesh buys many bones;
He that buys eggs buys many shells;
He that buys good ale buys nothing else.
17th Century English Proverb

Introduction

The novelist Graham Greene was no stranger to good beer. In fact, his family were well-established in the brewing business. As a young man, as was the experience of many adolescents, he found his first draught of the historic brown nectar to be an indecisive encounter. In his book *A Sort of Life* Greene recalled that he absolutely hated his first pint. But like most adventurers and seekers after truth, he had the courage to persevere, and reported the result of downing his second pint with the words "I enjoyed the taste with a pleasure that has never failed me since."

Greene's initiation into the wonderful world of ale will possess a familiar ring to thousands of thirsty men and women. In the coal fields surrounding Mansfield in Nottinghamshire, after a long perspiring shift in the gritty, black darkness at the coal face, there was no finer Holy Grail to occupy a pitman's imagination than that first pint of Mansfield's brew after a refreshing pithead bath. As men would hew the coal, a mile above them other men would brew their beer. Landlords and barmen recall a time when the staff of pubs in the vicinity of pits would keep their eyes on the clock. As the end of a shift approached, the top of the bar would be covered with pints of mild, bitter or mixed – a good publican knew every miner's order – ready for the thirsty colliery exodus about to descend upon their watering hole. Perhaps the only similar example of this we have today is the scramble to the theatre bar in the interlude for pre-ordered drinks. Beer was big business. Liquid food, refreshment, a palliative against a hard life.

This book has a similar connection to working men of another industry – the fishermen of Hull. In the 1970s, when the long-established Hull Brewery was first transformed into North Country Breweries, a subsidiary of the giant Northern Foods Plc., and then in 1984 absorbed yet again into Mansfield Brewery's growing territory, Hull's lovers of good ale eventually realised that they had nothing to worry about. Without Mansfield's timely involvement, they could well have suffered the total destruction of their taste buds if one of the more rapacious brewing giants had bid for their brewery. Soon the Mansfield logo was as welcome and well-known around the old port as Hull Brewery's old trademark had been, and despite the sun setting on Mansfield's original operation in its home town, the beer served up today by the new owners of the brand, Wolverhampton & Dudley Breweries, still possesses the character and flavour which made Mansfield famous.

The genesis of this book was the outcome of coincidence and chance which often result in new and unexpected pathways for a writer. After visiting France regularly every year since 1989 I was beginning to feel ashamed that I hadn't grasped much of the language beyond the ability to order a beer and a baguette. I therefore enrolled at Mansfield's West Nottinghamshire College for their weekly French night class which began in 2001. There, under the tutelage of the vivacious Helene Lovell (an unforeseen bonus of enrolment with which I feel sure most of my fellow male students would agree) I joined an enthusiastic group of Francophiles who were keen to finally grasp the nettle of speaking French. Each week I sat next to a tall, friendly gentleman who shared my often disruptive sense of humour. The life of a full-time writer is a solitary occupation, and this weekly exchange of eagerness for all things French between a group of like-minded people soon made Monday evenings the high-point of the week. It wasn't fully revealed that I earned my living as a writer until ego intervened and I proudly announced to some of my closer classmates in May 2002 that I had a new book out. This led to a brief exchange of information between us regarding our day jobs, wherein my immediate classmate revealed that he had spent the bulk of the last three decades of his working life as Mansfield Brewery Plc's Head of Personnel. After my own often bitter experiences of management in the corporate world prior to taking the plunge in 1997 to become a writer, I found John Else's retrospective passion for his role at the brewery and his residual warmth and close feelings for its staff very refreshing. Of course, being the company's Head of Personnel had put him in a unique position. He knew everyone. From the shop floor right up to the boardroom, he spoke of the workers as friends and gave thumbnail character sketches which would fascinate any writer with an ear for social history. Thus, when John outlined his plan to produce what is, in the main, an oral history of the last few decades of what was one of Mansfield's most popular and successful companies, and asked me if I was interested in becoming the literary shepherd for these diverse and fascinating yarns, I could only respond to such a compliment by saying yes. So this project owes a lot to John's sense of commitment and organisation. He provided the interviewees, set up all the meetings, and ferried me from hostelry to hostelry. It has been a totally enjoyable task.

The story of beer in Mansfield has already been the subject of a detailed and well-constructed history by Philip Bristow entitled *The Mansfield Brew,* (Navigator Publishing, Ringwood, Hants. 1976). Bristow's book remains a fascinating and comprehensive work which is packed with facts, figures and background political history. It encompasses all the main events, especially in the boardroom, from 1855 through to 1975. Without it the work you are now reading would have been immensely more difficult to complete. For a more basic overview, there is also Pat Etches and Donna Davis's *Memories From Mansfield's Industrial Past* (The Making It Industrial Heritage Trust Ltd. 2002). The nine pages in Pat and Donna's fascinating booklet devoted to brewing in Mansfield, compiled, as they were, in the period following the end of brewing in Littleworth and the take over of MBC by Wolverhampton & Dudley Breweries Plc provided more in the way of inspiration for style and content for this work, which picks up where Bristow's *Mansfield Brew* ends. It seems a strange coincidence, after living in Mansfield for a mere 20 years, that I should be telling the story of the Brewery which was partially involved in the demise of Hull Brewery. I was born and bred in Hull and although the first beer I ever became intoxicated on was Denmark's Tuborg, Hull's mild, amber and bitter ales formed the bedrock of my enthusiasm for beer. As temporary custodians of these brews, Northern Foods' North Country Breweries did an admirable job in the late 70s until they, in turn, were bought out by MBC. At least when I moved to Mansfield in 1987 I knew I was coming to a town where the quality of the beer would remain high.

This writer is therefore proud and honoured to be passed the baton so ably carried by those local historians who have gone before me. Looks like it's my round!

ROY BAINTON NOVEMBER 2006.

THE ALTERNATIVE BREW

TIMELINE:

1842 Mines Act – children under ten years old excluded from working underground.

1846 Sarah Baily (sister of William Baily) marries Francis Chadburn.

1849 First steam locomotive arrives in Mansfield.

1851 Queen Victoria opens the Great Exhibition at Crystal Palace in Hyde Park which is the first World Fair.

1854 The Company becomes established by
(1) William Baily of Mansfield, a gentleman farmer.
(2) Samuel Hage of Ollerton, a farmer.
(3) John Watson of Sheffield, a brewer.
In 1854 Mansfield was a malting centre, generally agricultural.

1855 Capacity in these early years reaches 200 barrels of beer per week. Early records show that fermenting vessel (F.V.) capacities were 4x25 barrels, 12 x 33 barrels, 3 x 50 barrels.

William Baily, Samuel Hage and John Watson become partners in the new venture and put their signatures to an agreement which includes the clause *'the Partners shall be true and just to each other'.*

Steam machinery begins to make industrial life slightly easier.

1856 As the Crimean War finishes, John Watson sells his shares in Mansfield Brewery for £450, leaving William Baily and Samuel Hage to run the business.

1863 The Partners set up in business with William Hage as Maltsters on 12th November under the name 'Hage and Baily' describing themselves as 'Wholesale Brewers of Mansfield'.

1864 The Great Northern Railway seek to extend their lines into the Nottinghamshire coal fields.

1868 Shacklock & Titley, Maltsters, set up their business in Mansfield. Henry Stephen Shacklock was the cashier at Mansfield Brewery, and Addison Titley, the other partner, was the proprietor of a successful flax spinning business in Leeds and chairman of the Liverpool, London & Globe Insurance Company, a Justice of the Peace and a director of the Midland Railway Company – not bad for a young man of just 20!

1871 The Mansfield – Southwell Railway opens. The railway passes through a steep cutting at Python Hill where the family home of William Baily stood – the bridge across the cutting becomes known as Baily's Bridge.

1873 Addison Titley is admitted into the Mansfield Brewery partnership for an investment of £2,000 – "a sixth of the ascertained goodwill".

The Mansfield Brewery premises are now insured for £4,130. The rent paid by Hage and Baily is £377 per annum. Tiltey is paid £150 per annum.

1874 On January 5th, William Baily dies.

1875 Opening of Mansfield-Worksop Railway following the completion of Mansfield's Church Street Viaduct.

William Jackson Chadburn (son of Francis Chadburn and nephew of William Baily) joined the company as a Brewer's Clerk. He became a partner in 1885.

1877 Shacklock sets up as a maltster again, this time with William Jackson Chadburn with a capital of £2,100. The business is carried on in Mansfield in the Eclipse Inn Yard, Westgate, Brick Lane West and Ratcliffe Gate.

Pleasley Pit is sunk.

1878 Brewery partnership on March 1st sees Addison Titley owning one half of the company, William Edward Baily (the late William's heir) owns one third, and his uncle, John Baily (William's elder brother) the remaining one sixth share.

1879 Mansfield Brewery profits under the supervision of Addison Titley begin to soar, increasing six fold in 1879 to £6,000.

The brewery owns 19 public houses and leases a further 16.

1880 New railway lines are opened further connecting the Mansfield area with the main through-lines through Nottingham.

At the Martinmas Hiring Fairs and the Nottingham Goose Fairs the Mansfield Brew is the favourite, with many thousands of gallons being sold.

1881 Foundation of Nottinghamshire Miners' Federation.

1884 William Edward Baily, now married to the daughter of the landlord of the White Hart, celebrates the birth of their daughter, Gertrude Sarah.

Addison Titley, after twelve years of running the brewery, is considering his retirement and agrees to dispose of most of his shares in the company.

1885 A new Mansfield Brewery partnership is established with William Edward Baily, Henry Stephen Shacklock, William Jackson Chadburn and a new partner, from The Lounge in Albert Street, Mansfield, James Handley Hopewell. Hopewell brings a new expertise to the company's business gained from his experience in a successful wines and spirits business.

William Jackson Chadburn's first son, Francis Ludlow Chadburn is born 13th September.

Over the next four years, the Brewery add to their pubs at an impressive rate by buying:

The Old Eclipse, Westgate	£3,350
The Bowl in Hand, Leeming Street	£2,000
Speed The Plough, Sutton-in-Ashfield	£1,250
Peacock Inn, Hucknall	£1,375
Forest Tavern, Annesley Woodhouse	£1,020
Dial Hotel, Market Place, Mansfield	£3,000
Stag & Pheasant, Leeming Street, Mfd.	£2,200
Railway Inn at Pinxton Station	£2,000
Queen's Head Inn, Mansfield	£2,400
Ram Inn, opposite the brewery	£ 750
Wagon & Coals, Ratcliffe Gate, Mfd.	£1,680
Forest Tavern, Skegby	£ 870
Angel Inn, Clowne	£1,675
New Inn, Mansfield Woodhouse	£1,380
Kings Head Inn, Mansfield	£1,600
Plough Inn, Ollerton	£ 800
Old White Bear, Mansfield	£1,610
Rifle Volunteer, Skegby	£1,161
Queen's Head Inn, Alfreton	£1,100
Yew Tree Inn, Mansfield	£ 970
Old Trooper Inn, Sutton-in-Ashfield	£2,100

1887 Queen Victoria celebrates her Jubilee.

1888 William Jackson Chadburn's second son, Claude William Chadburn, born 7th February.

1889 New large collieries are developed at Kirkby Summit, Bentinck, Cresswell, Warsop and South Normanton.

1891 Mansfield is granted a Municipal Charter on the 14th July and 30,000 people attend the ceremonies.

The brewery buys three more pubs.

1892 Another eight pubs are purchased in the year when the first telephone is installed in Mansfield. Mansfield Brewery were among the 22 subscribers enrolled by 1895. William Jackson Chadburn is reluctant to use the phone.

Keir Hardie speaks at Bestwood.

1893 This was the year of the coal mining lock-out but it did not prevent Mansfield Brewery from purchasing a further pub, the Apollo Tavern in Sutton-in-Ashfield.

Keir Hardie, founder of the Labour Party, has become MP for West Ham South.

1894 Over a four year period, ten pubs are added to the growing Mansfield Brewery empire.

1899 William Jackson Chadburn becomes Mayor of Mansfield.

Henry Stephen Shacklock dies and his shares are bought by William Edward Baily.

1900 Mansfield Brewery employs their first accountant.

1901 Queen Victoria dies at Osborne House aged 81. Edward VII's accession to the throne.

James Hopewell dies and his shares are bought by William Jackson Chadburn.

The company is now owned by three families, Baily, Chadburn and Titley.

1903 Co-founder Baily's son, William Edward, dies May 26th. His shares are left on trust to a Penzance solicitor, George Bodilly and William Jackson Chadburn. His widow and daughter, Gertrude, move to Worthing.

The company plan their strategy around the development of the coal fields, which dramatically increases the local beer consumption.

The smaller, old 'cottage' pubs, are swamped with drinkers, so re-building becomes a policy in the Kirkby, Sutton, Huthwaite, Pleasley and Woodhouse areas.

1904 The Brewery partnership is now as follows: Trustees of William Edward Baily now manage half the shares, William Jackson Chadburn owns one third, with the sons of Addison Titley, Edward and Walter, sharing the remaining sixth.

1905 Gertrude Sarah Baily became 21 years old in May. Increased demand causes some difficulties regarding biological control, so a decision is made on the building of a new brewhouse.

Property is sold to help to raise the £20,000 for this. It requires a lot of faith and courage, as everyone says the plan is too bold.

1906 Claude W. Chadburn joins the company at the age of 19.

Harriet Jarvis, a widow, moves into the Ram Inn opposite the Brewery with her children Harry (7) and Jessie (4).

1907 New brewhouse started. Trade was 17,000 barrels per annum.

1908 Stone fermenting vessels were being replaced by slate. Trade was 20,000 barrels per annum.

1910 Edward Titley marries Harriet Jarvis.
Edward VII dies of pneumonia and is succeeded by George V.

The land to the rear of the Brewery is sold to the Great Central Railway Company, which halts further development.

1912 The Brewery Partnership at this time sees the trustees of William Edward Baily owning one half of the company. William Jackson reduces his holding to one sixth by giving his other sixth to his sons Francis Ludlow Chadburn and Claude William Chadburn.

Edward Titley increases his shareholding to one sixth.

William Jackson Chadburn moves from Grove House into Nettleworth Manor.
Francis Ludlow Chadburn marries Kate Elizabeth Adams and moves to Ollerton House.

Claude William Chadburn marries Ethel Pegler from Darley Abbey and moves to Woodside on Chesterfield Road, Mansfield.

Coal Mines Minimum Wages Act is passed in response to the first national coal strike.

1914 Outbreak of World War I.

1915 Gertrude Sarah Baily marries Frank Munro Thomas, a Junior Master at St. Dunstan's Prep School for Boys in Worthing. He joined the Army soon after his marriage and reached the rank of Captain.

At Clipstone Army Camp (near Mansfield) 60,000 thirsty soldiers are stationed.

1916 Frank and Gertrude Thomas celebrate the birth of their son, Frank Bruce Baily Thomas.

1918 Sydney Deakin joins the company as the only Brewer (their first trained brewer). He makes an early improvement by converting copper boiling from fire to steam.

World War I comes to an end.

Claude W. Chadburn returned from active service and moved to Beech Hill.

1919 To increase capacity fermenting vessels were bought from the closed brewery at Pocklington in Yorkshire – 4 x 50 barrels.

Francis Ludlow and Kate Elizabeth Chadburn celebrate the birth of their son Robert William. (Known throughout his life as Robin)

1920 Pits sunk at Clipstone and Thoresby.

Commencement of Welfare Levy for miners.

It became apparent that Frank Baily Thomas was mentally handicapped.

1922 Following the death, at the age of 37, of Francis Ludlow Chadburn on April 5th, his father William Jackson Chadburn dies May.

First Labour majority in the House of Commons under Labour leader Ramsay MacDonald.

1923 Claude Chadburn takes over one third of the company and becomes Chairman.

Edward Titley dies aged 49.
Extension to the fermenting room.

1924 Over the next six years, public house building and re-building takes place such as the re-built Victoria and the new Black Bull, Rufford, and the Woolpack.

Gertrude Sarah (Baily) Thomas dies. Her closest friend, Miss Molly White, is appointed as guardian to Frank Baily Thomas, now 7 years old.

1925 Mansfield Brewery ceases to be a partnership and becomes a Limited Company, Mansfield Brewery Ltd. with Claude Chadburn as Chairman. The directors are: Claude Chadburn, Harry Jarvis (Edward Titley's step-son) and Captain Munro Thomas.

Claude Chadburn buys Papplewick Hall.

1926 Miners go on strike May 1st. followed by the General Strike, called by the TUC in support of the miners.

Edward Deakin joins as a pupil brewer then leaves, to re-join in 1928.

1929 Francis Chadburn's widow, Kate Elizabeth, dies February 8th aged 47.

1930 Abolition of the 20 mile per hour speed limit.

1931 Edward Titley's step-son, Harry Jarvis, marries.

During the depression years further development is halted, with the low spot being reached in February 1931, when only three hundred barrels are produced in one week.

1934 30 mile speed limit in towns imposed.

1935 The company is converted to a Public Company, and acquires Chesterfield Brewery which at that time is only producing 100 barrels per week, yet has nearly 100 houses.

Group pre-tax profits £56,000.

1936 Fred Hensleigh commences employment in April.

George V dies. Edward VIII abdicates.
George VI's accession to the throne.

J. B. Jenkins joins the Brewery as accountant June 15th.

1937 The King's Coronation.

The next two years sees Fermenting capacity further extended with a replacement of the old Pocklington fermenting vessels. They cost £150 each (compared with, for example, a 250 barrel F.V. in 1977 at approximately £20,000).

There is further investment in public houses such as Pleasley's White Swan and Mansfield's Oak Tree and Sir John Cockle.

1938 John Ludlow Chadburn, son of Claude, is killed at Newstead in a car accident.

J. B. Jenkins succeeds Tommy Gilbert as Company Secretary April 5th.
T. P. Wood & Co., wine and spirits section of Chesterfield Brewery, finally comes to Mansfield.

Land adjoining the Brewery is purchased from the electricity department to the lasting regret of the East Midlands Electricity Board.

The area now becomes the spirit and inflammable paint store, loading area and boiler house.

In December, Philip A. Baker joins the company as pupil brewer.

1939 Sydney Deakin becomes a director of the company.

The last five horses are sold; at its peak the stables had 25 horses.

Britain declares war on Germany.

1942 Alan William Chadburn, son of Claude, is killed at Tobruk on June 6th.

September 5th Dorothy Fowkes (whose family has connections to the Metal Box Company) marries Robin W. Chadburn.

1944 4/5 acres of land on Great Central road are purchased for a bottling store and land is also bought for a boiler house.

1945 National Union of Mineworkers formed on January 1st.

Second World War ends.

1946 Gordon Kendall commences employment in January.

Edward Deakin and J. B. Jenkins become directors June 18th.

1947 Following distinguished active war service, Philip Baker who had worked as a pupil brewer before the conflict, re-joins the company.

Sydney Deakin retires.

Coal Industry nationalised.

1948 Railways nationalised.

National Health Service formed.

November 5th Robin and Dorothy ('Bo') Chadburn celebrate the birth of their son, Richard Fowkes.

1949 Gas is nationalised.

1950 Steel is Nationalised.

Additional mash tun, new bottling hall and new boiler house.

Albert Gregory commences employment on May 1st.

The Titley interest fades out.

During the coming decade, there is a huge upsurge in the company's business.

1951 Harry Jarvis dies, April 29th

Madge Branfield commences her employment with the Brewery in the accounts department.

1952 King George VI dies.

Queen Elizabeth II's accession to the throne.

1953 The Queen's Coronation.

1955 Hornby's soft drinks business is acquired and transferred to the Brewery. Hornby's greatest assets are its Public Houses. The best site was reputed to be Savage's Vaults adjoining British Home Stores and The Old Eclipse. Eventually the latter was sold to British Home Stores. The other pubs were:

The Plough, Farnsfield
The Crown, Southwell,
The Vine, Chapel St. Leonards,
Ravensdale, Mansfield.

1956 Edward Deakin and J. B. Jenkins become joint Managing Directors August 1st.
Philip Baker becomes Head Brewer and Fred Gleadell becomes Company Secretary.

1957 August 12th Robin W. Chadburn and Jocelyn 'Jock' Nangle (Trust solicitor for the Baily Trust) joined as Directors.

Michael Parsons joins as second Brewer.

1958 Slate fermenting vessels replaced with stainless steel.

1959 Sydney Deakin dies July 20th.

1960 Bright Beer introduced on experimental basis.

Brian Oldham joins the company as a cooper.

1961 Robin Chadburn becomes deputy Chairman.

Company pre-tax profits £500,000.

John Goulden commences his employment with Hull Brewery in March.

1962 John Bartle commences employment in July.

David B. S. Weston appointed to the board November 22nd.

Number 3 fermenting room completed.

1963 Claude W. Chadburn became President of the company and Robin W. Chadburn becomes Chairman.

Captain Munro Thomas retires.

David Weston killed in a riding accident at Papplewick.

Tony Scruby joined the company as joint Managing Director designate.

Christmas sees the introduction of the 7-pint can.

1964 Alice Donbavand commences her employment.

1965 W. N. Donbavand (Bill) commences employment in July.

J. B. Jenkins retires September 30th.

Tony Scruby becomes joint Managing Director with Edward Deakin.

New tenancy agreement introduced and sales company formed.

1966 David T. N. Tyne joined the company as Chief Accountant.
Completion of stainless steel conversion of fermenting vessels and number 4 fermenting room.

1967 New bulk beer hall operating.

Meadow foundry site purchased.

1968 Jean Elliott commences employment in September.

1970 Barbara Brown commences employment in March.

First long service awards made.

Philip Baker joins board November 18th.

Captain Munro Thomas dies.

1971 The Railway having being abandoned, the land reverts to Brewery ownership at a cost of £20,000.

In February Decimal currency introduced in UK.

In July, the Red Lion in Birchover is sold to sitting tenant.

Pre-tax profits are £1,065,000.

Edward Deakin retires September 30th.

Tony Scruby now sole Managing Director.

Michael Parsons becomes Head Brewer October 1st.

The Baily Thomas Charitable and Provident Funds are founded, the first trustee action is noted in December, the work being carried out by Jock Nangle.

1972 Miners' strike January and February.

Extension of cold storage in bulk beer hall, completed in December, and the end of wooden casks.

1973 Stephen (Sam) Sills commenced employment in April in the wines and spirits department.

With the occasion of the Royal Wedding between Princess Anne and Mark Phillips, the Brewery decide to brew a special beer in September.

Hugh W. Wilson joins the company in the Properties department in October.

New building completed for cask filling, wines and spirits and bulk road tankers. This saw the last of cask-conditioned beer for some years and wooden casks forever, with the new kegging plant fully operational in October. A farewell party is held to drink the last 18 gallons in the cellars.

1974 In May the board decided to use TV advertising for the first time.

In August it was decided to cease malting.

New canteen and changing rooms completed.

Wines & Spirits transferred to new building.

Richard H. Umbers joined the company as Sales Director in October.

First investment in R. L. Jones & Co. Ltd. (Later to become Mandora soft drinks).

1975 The Brewery's first ever strike lasts seven days in May.

In June John Else joins the company as Training Manager, on a split contract with Hardy's & Hanson's Kimberley Brewery.

The first bulk lager tanks are added to the bulk beer hall.

Marksman Lager is introduced.
A new laboratory is opened in September.

At the end of September 75 years of bottling beer in Mansfield comes to a close. The bottling department becomes a warehouse plus a small area for canning.

1976 In April the first conical F.Vs in operation at Mansfield.

Ceased manufacturing of soft drinks in June.

New offices and reception opened.

Marksman Lager on sale to all tenants.

Seven new pubs opened during the year.

Sid Tyler retires after 49 years. Both his father and mother had worked for the Brewery and Sid followed the family tradition.

1977 Claude W. Chadburn dies peacefully at home at Papplewick on February 27th.

In April, the company's house magazine, Marksman, is published for the first time.

In July, a history of the company is published, Philip Bristow's The Mansfield Brew.

Terry Johnson commences employment in July.

Pre-tax profits £2,459,000.

David Tyne joins the Main Board August 1st.

In August 16 oz. cans of bitter and lager are introduced.

R. L. Jones & Co., a soft drinks business, under the name Mandora, is acquired.

Mansfield Inns is formed and will be run by Colin Lowther, who joins MBC on September 19th as a divisional director.

Mike Cleator and Bob O'Meara commence employment in October.

In November it is reported that the Coal Board Pension Scheme has purchased 6% of MBC equity.

1978 Brewery now has only bulk and kegs, as bottled products are transferred to Mandora.

John Else becomes full time at Mansfield in May.

Take over bid of £18 million in cash by Mansfield Brewery failed to secure Shipstones of Nottingham.

Cellars reception area opens July 3rd.
The Company Secretary, Fred Gleadell, retires, to be replaced by Mark Phillips.

1979 May: The sad death of Graham Reason (cellar installation) from an asthma attack on return from Supervisor's annual outing.

'Death In Service': Graham Wilson, who had been retired only a few weeks, dies in July at 65 after 44 years' service with the company.

Tony Scruby retires and John F. Hings takes over as Managing Director.

Richard F. Chadburn appointed to the board as director in December.

1980 January: Finance Director David T. N. Tyne leaves the company, and Chris Rainsford transfers from Mandora to take Tyne's place.

April: Alec R. Welham appointed as Group Financial Controller.

May: Richard Umbers leaves and Dennis Foster is appointed as Free Trade director.

September: The Brewery buys T. W. Beach & Sons, a soft drinks business in Evesham.

December: The soft drinks side of the company is consolidated as Mandora (UK) Ltd.

Mandora's Managing Director Tony Morton joins MBC's main board.

1981 Re-openings after re-furbishment of Mansfield's Wheatsheaf, Swanwick's Boot & Slipper Inn and the White Lion at Blidworth.

April: Philip Baker retires.
Michael Parsons becomes Brewing Director.

June: Mark Phillips changes role from Company Secretary to become Tenanted Trade Director.

Richard Chadburn becomes Company Secretary.
Richard M. J. Meadows joins company as Brewing Production Manager.

July: The Royal Wedding of the Prince of Wales to Princess Diana. The brewery issues a gift pack of drink and glasses to every employee.

August: Profits for the year ending April 1981 are up by 32% with turnover increased by 44%.

September: Colin Lowther leaves the company to be replaced by Derek Mapp who is to be Retail Director of Mansfield Inns.

The Van Dyk Hotel is re-opened after a six month closure for complete internal re-modelling and extensive external work at a cost of £350,000.

The 'Mandora Rocket' is erected.
Financial Director Chris Rainsford develops the company's new computer department.

Ron Allen and Alf Ansell celebrate 50 years service with the company and are presented with gifts by the Chairman, Robin Chadburn.

October: A new public house, The Sawmill, Newark, is opened, the cost of the project £304,253.00

November: Change of name to Mansfield Brewery plc.

The second saucy Mandora calendar is a great success.

A new wine bar is opened in the cellars of The Oak Tree Inn on Southwell Road, Mansfield.

December: The announcement is made that Mansfield Brewery is to build a new brewhouse on the old loading stage site. The existing brewhouse dates from between 1906 – 1950.

Robin Chadburn carves the turkey on Christmas Eve at the Brewery Christmas party at The Cellars.

1982 January: The legendary Albert Gregory retires after 30 years. David Rhodes takes over Albert's role.

Derek Evans joins Mansfield Brewery plc as Wines & Spirits Sales Executive.

Ron Allen retires after 51 years' service. Ron was the first beneficiary of a recently introduced non-contributory pension scheme and this made a significant difference to Ron's retirement.

February: Marksman, the new, 'continental tasting' lager is launched with a high-profile advertising and marketing campaign.

Ted 'Dramatic' Taylor, (inventory & stationery control) who has served with the Brewery for 22 years, scores a hit as the policeman Sergeant Rough in the Victorian thriller, 'Gaslight', at the Mansfield Civic Theatre.

March: Mobile Chest X-Ray unit, the first to visit the Brewery and Mandora.
Harry Butler retires after 33 years.

The Lord Mayor of Sheffield, Councillor Mrs. Enid Hattersley, visits the brewery.

April: Return to cask conditioned beer with 4x and the kegging plant is extended, 4 pint canning ceases and 2 litre PET (Poly Ethylene Terephthalate) bottling begins.

Brenda Walker becomes the company's first-ever Pensions Co-ordinator.

May: Tony Scruby retires from Chair of Soft Drinks Board of Directors.

June: Alf Ansell, Supervisor of cask filling department retires after 51 years.

Brian Oldham moves to cask filling and a new Supervisor, Brian Lucas, is appointed.

Desford Dowty Brass Band of Whitwick, North Leicestershire, win the Granada TV Band of The Year Award and the BBC's Best of Brass Award, attributing their success to 'a regular intake' of Mansfield Bitter.

July: Decision taken to invest £1.5 million to manufacture PET bottles at Mandora Mansfield and PVC bottles at Mandora Evesham.

September: Mr. & Mrs. Robin Chadburn's Ruby Wedding: all employees and their spouses were entertained at Sudbrook Hill Farm on September 4th During the celebrations 69 employees receive long service awards.

Wendy Courtman, Mandora's Assistant Financial Accountant, receives her Duke of Edinburgh Gold Award at Buckingham Palace.

October: The new brewhouse building plans are approved.

November: Contracts are signed for the construction of the new brewhouse with a budget of £3,914,250.

December: Ray Mellors retires after 47 years.

The *South China News* reports the story of a Mansfield man admitted to Nottingham City Hospital with serious burns whose only request was for a pint of Mansfield Bitter. The Brewery duly sent the patient two dozen cans.

1983 January: Ian Tomlinson, son of Arthur Tomlinson

(properties manager) is appointed as Operations Manager of Mansfield Inns.

February: Les Yates, Technical Services Supervisor, retires after 48 years service. Ian Boucher takes over Les' role.

March: Chris Rainsford is appointed Managing Director of Mandora (UK) Ltd.

New conical fermenters are installed.

April: Marksman Lager is voted Supreme Champion Draught Lager at 1983 Brewex.

Ron Kirk joins the brewery as Group Financial Director.

July: Implementation of the costly Mansfield Brewery Passport promotion.
August: After 21 colourful years managing the wines and spirits department Harry Ridley retires.

September: Ken Perkins retires after almost 50 years with the soft drinks operation.

Launch of beer advertising campaign featuring Ronald Reagan (later to be followed by Mikhail Gorbachev and Muhammad Ali).

October: Neil Tomlinson (Arthur's other son) appointed as Property Supervisor.

New Brewhouse topping-out ceremony.

1984 January: The decision is taken to investigate the possibility of the American market for Mansfield Beers.

Installation of 1,000 barrel conditioning casks in beer process, and 40 head PET filter. I litre PET started.

March: Brian Roberts joins company as General Manager, Marketing.

Miner's Strike begins which will last almost 12 months.

April: Fred Hensleigh retires after 48 years service.

May: J. B. Jenkins, previous joint Managing Director, dies.

July: The new brewhouse opens. There are open days for employees and their families, and at the end of the month a special opening for local dignitaries and shareholders. The official opening ceremony takes place in August, with the total bill for these occasions reaching £59,000.

Brian Oldham is appointed Supervisor of the new brewhouse.

October: A bonus / share option scheme for employees is introduced.

November: Dennis Bennett retires after 30 years.
December: George Sirdfield retires after 37 years and Walter Murphy after 30 years.

1985 March: A bar fittings and supplies company,

William G. Bowler & Son, is acquired.

Opening of refurbished Mansfield Hide & Skin factory as an Enterprise Project Centre by Tom King, the Minister of State for Employment.

The Enterprise Centre is managed by Ian Boucher who is on secondment from his role as Technical Services Supervisor.

MBC directors visit the USA to research markets for Mansfield Brewery products.

On March 30th Chairman Robin Chadburn attended the official opening ceremony, by Dennis Thatcher, of the new Territorial Army Headquarters.

Doris Watson retires after 36 years.

April: Derek Mapp appointed Managing Director for Mansfield Inns.

June: Filming of the ‘Not Much Matches Mansfield’ TV adverts.

July: Richard Chadburn appointed as Managing Director of Mansfield Brewery Trading Ltd.

Marksman Lager scores again as Supreme Champion Draught Lager at the Brewing Industry International Awards Exhibition 1985.

North Country Brewery, based in Hull, is taken over for £42 million.

October: The Frampton Village Cider Company is formed jointly with Hazlewoods.

The decision is taken to distribute a copy of the annual accounts to all employees.

November: The staff of Mansfield Inns move to leasehold premises on Rosemary Street.

2½ million pint storage expansion completed.

3 litre PET started.

It was agreed that Lambert, Parker and Gaines would be the name used for the wines and spirits department.

American launch of bottled Marksman lager.

During the year John Else pulled off a training coup by organising a presentation by the charismatic John Garnett, Head of the Industrial Society. The theme of the presentation was staff communication. The audience included all the Mansfield Brewery Group Directors and Senior Managers. John Garnett's presentation had quite an impact on the gathering, with a passion and commitment second to none. In December, a few training sessions later, Team Briefing commenced.

1986 January: Alan Moss joined Mandora (UK) Ltd. as Marketing Director following the resignation of Richard M. Lewis.

Hull employees and their families visit Mansfield Brewery.

February: Aged 63, Eamon Austin O'Donoghue dies from a massive heart attack whilst in Chesterfield Hospital. Austin, a Tenanted Trade Manager, was an inspiration to all who worked with him and will be sadly missed.

Chris H. Downton joined the company as Company Secretary.

The purchase from the Official Receiver of Rimark of Wellingborough for £325,000 for assets and goodwill and £94,000 for stock was concluded.
Arrangements were made to execute a number of export contracts and other commitments.

Consideration was given for the removal of plant, etc. from Wellingborough.
The acquisition of Rimark gave the company the St. Clements soft drinks brand.

April: Arthur Tomlinson, Estates Manager, retires after 24 years' service.
Kevin Hall is appointed as Building Supervisor in the Properties Department.

May: Property divestment: Camerons pay Mansfield Brewery £13 million for the unwanted North Country Brewery pubs together with a few Mansfield houses.

Frank Grady retires after 46 years.

July: The brewing industry was once again referred to the Monopolies and Mergers Commission by the Office of Fair Trading. This resulted in a constant work burden at Mansfield in relating facts and figures to particular criteria. Work continued through 1986, 87 and 88. The final result changed the face of the brewing industry out of all recognition.

August: Edward Deakin dies. He had served the company from 1928 to 1971.

September: Chairman Robin Chadburn is appointed as Chairman of the Brewer's Society Public Relations Committee.

October: Cask warehouse extension operational.

Michael Parsons dies. He had been Brewing Director since 1981 and his service with the company began in 1957. His qualities and skills are much missed at Mansfield Brewery.

November: Harry Battersby, Area Sales Manager, dies in a tragic car accident after stopping on the motorway to help others. Aged 62, he had been with the company since 1967.

Charles Kerridge retires after almost 40 years service.

Steve Holt joins the Brewery trading company as Marketing and Commercial Manager.

Richard Meadows is promoted to Brewing Director.

1987 January: Decision to move Properties department and H. W. Wilson to Dame Flogan Street in Mansfield.

March: Madge Branfield retires from the company after 36 years of service, having joined in January 1951. She had been in charge of the Brewery Reception for 20 years.

Brian Nicholson, Chairman of the Manpower Services Commission visits the Mansfield Brewery Enterprise Project Centre.

Tony Morton left the company.

May: Decision to change the name of 4X Strong Beer to Old Baily Strong Beer, and to brew a 1035 product called Riding bitter.

St. Clements brand gets £2 million marketing boost.

The nationwide TV campaign starts in mid-May.

Decision is taken to develop Harpings Road depot, Hull, to allow evacuation of the old Hull Brewery site in Silvester Street.

June: Appointment of Paul D. Cooper as Divisional Director.

Chinese trade delegation visits the Brewery.

July: John Hings proposed the introduction of Fosters as an internationally promoted lager brand.
Both Fosters and Red Stripe were introduced in September.

Ken Kinch, assistant Company Secretary retires after 41 years' service.

Jim Wain, Transport Supervisor, retires after 50 years' service, having joined the company on August 10th 1937. Jack Revill retired on the same day after serving 50 years, and had joined on the same day as Jim Wain, following him into the Transport department. Roy Bailey takes over from Jim Wain and Roy Thorpe becomes Warehouse Supervisor.

August: Tom King, now Secretary of State for Northern Ireland, was most interested in the additive free products on display at the N.I. Food Exhibition, particularly the St. Clements products.
The company buys the UK franchise for Rainbow Carpet Cleaning and Dyeing for £500,000.

September: Brewery's biggest ever marketing spend of £2 million to support 'The Taste of '87'.

October: It was agreed to open up health screening to a wider range of employees.

Barbara Toon retires after selling Mansfield products for 21 years.

November: Steve Holt, Marketing Director, remarkably escapes the tragic fire on the underground at King's Cross in London, much to his colleagues' relief at the brewery.

David Tait, Mandora laboratory assistant, receives his Duke of Edinburgh Gold Award at St. James' Palace.

December: Mandora St. Clements was sold to A. G. Barr & Co, Scottish manufacturers of Irn Bru, completion date February 1988.

During the year, installed 62 head PET filler. Half litre PET started.

A number of pubs have major refurbishments as part of a committee £8 million capital investment programme

1988 January: Sam Sills leaves the Brewery to join IDV as Croft Port Salesman.

February: Ken Evison becomes the first Mansfield Brewery Licensee to be elected Chairman of The Licensed Victuallers' National Homes Charity.

April: Purchases the franchise for Bonanza, an American catering concept.

John F. Hings tendered his resignation as Group Managing Director.

Ron L. J. Kirk is appointed acting Managing Director.

The Spring of 1988 sees the launch of the company's new Medical Screening Programme called Healthline. Details are published outlining the retention of a Company Doctor and the extensive new coverage of additional employees.

Mansfield MP Alan Meale and Hull MP John Prescott visit the Brewery.

July: Mansfield Brewery plc purchases Hazlewood's share of the Frampton Village Cider Company, which becomes wholly-owned by Mansfield Brewery plc.

Franchises obtained for Berni Restaurants.
The Rushley, due to re-open in November, will be the first Berni franchise in Britain.

September: Robin W. Chadburn's request to retire in March 1989 is accepted. Geoffrey C. Kent is appointed Deputy Chairman in a non-executive capacity until 31st March 1989 with a view to his becoming Chairman at that date.

Ron L. J. Kirk is appointed Group Managing Director.

Derek Mapp and Hugh W. Wilson both appointed executive Directors.

A new concept, Landlord's Table Restaurants, is introduced.

Chris H. Downton leaves the company, but generously agrees to act whilst a replacement is found.

1989 January: The Monopolies and Mergers Commission regulations take full effect on the1st.

March: Robin Chadburn retires from the company on the 31st and the following day is appointed President for life. Mr. Nangle pays tribute to Robin's contribution to the company's success, and recalls that he and Robin had both joined the company together in 1957.

Jock Nangle also resigns as a Director with effect from March 31st, and the following day becomes Consultant to the company. Robin Chadburn reciprocates his colleague's accolades and congratulates Mr. Nangle on his contribution to the company.

April: A civic banquet is held at Mansfield's Civic Centre to mark Robin Chadburn's retirement. It is attended by civic dignitaries, members of the Brewers' Society and most of the UK's breweries are represented. Many personal friends also attend.
The banquet was organised unbeknown to Robin. To maintain total secrecy the project was referred to as 'Golden Eagle' (a sign of the times) . The staff involved in Golden Eagle were Project Chairman John Else, Cheryl Higginbotham, Tracey Stocks and Terry Wiseman. On the night Terry stood in for Michael Aspel during the 'This Is Your Life' spot. Syd Lawrence conducted his band and played into the small hours.

May: Between May 3rd-5th farewell parties for Robin Chadburn are held at Portland Training College Mansfield and are attended by all the workforce and their families, including pensioners.

On May 6th three coaches of employees and families are taken to the Bonanza Restaurant in Bayswater.

June: Colin R. Stump is appointed Marketing Director along with John G. Hare as Company Secretary.

July: A copy of the Company Philosophy was sent to each employee by the Chairman Geoffrey Kent.

September: Riding Best Bitter (keg) and Riding Traditional Mild are introduced.

November: Eric Jackson retires after 36 years service.

December: G. Paul Handley commenced as Finance Director and Mr. Vic Steel, Executive Director of Kingfisher plc was appointed non-executive director of Mansfield Brewery plc.

1990 February: Eileen Ashley, Secretary to the Chairman, retires after nearly 20 years of service, during which time she had served as Secretary to Robin Chadburn, Tony Scruby, John Hings and latterly Geoffrey Kent.

June: Riding Bitter is launched on Yorkshire Television, a continuing TV campaign which is responsible for Riding's growth and success.

Bonanza restaurant opens in Blackpool.

New cask racking plant installed.

July: The Enterprise Centre workshops are handed over to the Mansfield, Sutton and Kirkby Enterprise Partnership.

Ian Boucher transfers to the Wine & Spirits department as Warehouse Manager.

Appointment of John Else and Ian Tomlinson as Divisional Directors.

During the year Rainbow Carpet cleaning business is sold to a management buy-out team.

The Frampton Village Cider business is sold to Bulmers.

Berni Restaurant and Hotel franchise is discontinued.

Bonanza Restaurant business closed.

1991 Alice Donbavand retires after 26 years.

July: sees the opening of the new £3million Head Office based opposite the Brewery in Littleworth.

Low Profits: Managing Director Ron Kirk outlines the affect the recession is having on profits:
'Profits this year have been doubly hit as we also suffered one-off costs arising from £300,000 loss of profits because of the appalling weather conditions in the ten days following the snow storms in early December last year when the Brewery was forced to close, and the majority of our pubs in Nottinghamshire ceased to trade.'

Trading profit was down 5.9% at £8m.

The costs of exiting Rainbow Carpets and Bonanza Restaurants also contributes to a profits downturn. This strategy led by Geoffrey Kent took Mansfield Brewery back to doing what it knew best.

October: Bill McCosh joins the company as Managing Director.

Peter Fairey, landlord of the Robin Hood Baslow is awarded a British Empire Medal for his work before and during the Gulf War. Peter retired from the Navy in September.

29 John Smiths' pubs are purchased from Courage (managed houses).

Grolsch Lager is added to a now very competitive portfolio of draught products.

PET lines are realigned with Pozimat unscrambler. Now 3 litre, 2 ¼ and 2 litre sizes only.

Gordon Stocks retires after 30 years.

During the year Boaler Bar Supplies business sold to Autobar.

1992 February: 88 John Smiths Pubs acquired from Courage.

6 Whitbread pubs acquired.

April: Riding Bitter ads launched on both Central Television and YTV, establishing the brand as a major TV advertiser.

More than £1.5 million is being invested in pubs throughout the estate in the first half of the year.

July: Events Week raises a record £46,000 for local charities.

August: Old Baily is launched in a can.
Profits back on track at £11 million – up 36.8%.

September: On the 23rd the Chairman, Geoffrey Kent, dies, undergoing a routine operation.

During the year the refrigeration plant is expanded at a cost of £1.5 million to meet the new level of demand where brewings peak at 12,000 barrels per week.

Canning at 80,000 barrels per annum is moved from the Co-op at Sutton Coldfield to Ben Shaws at Pontefract.

The Keg Plant now moves to 3 shift working.

1993 March: Sir David White joins the company as Chairman on the 9th.

11 pubs are bought from Scottish & Newcastle on the 25th.

May: McEwan's lager (S&N) replaces Marksman lager within tied and free trade outlets, but is still available in cans and PET.

John Hubbard receives the British Institute of Innkeeping's Martell Gold Award for the highest membership entrance examination marks in the country.

June: Brewery Magazine changes its name from The Marksman to The Brewery Independent.

Mansfield Beers are exported to Russia. Bob O'Meara said "We've been so successful it now looks as if Mr. Yeltsin will soon have the chance to try our brews."

Profits are up and the amount paid into the employee profit share scheme rose to £½million.
July: Events Week set yet another all time record by raising £50,000.

Albert Mitchell retires after 39 years and Ray Crocock after 49 years' service.

August: Mansfield Cask is re-launched after an absence of a number of years.

September: Richard F. Chadburn and Jocelyn (Jos) Nangle follow in their fathers' footsteps and become Trustees of the Baily Thomas Funds.

November: John Simpson retires after 29 years.

By the end of the year another £2.5 million has been spent on new pubs.

Peter Witham, George Powell and Joe Crosby retire after serving 101 years between them.

1994 The launch of the Serving Quality Programme of continuous service improvement, a programme which has involved a £½ million investment.

January: Mansfield MP Alan Meale and Hull MP John Prescott, both members of the Parliamentary Beer Club, make a return visit to the Brewery.

February: Riding Bitter voted Supreme Champion at Brewing Exhibition. Mansfield Cask wins a Gold Medal, Mansfield Mild Keg wins a Bronze.

The Publican Awards: Mansfield Inns is voted Best Pub Chain.

The Grouse & Claret, Rowsley made it to the finals of The Catering Pub of The Year.

During the year the Crispin Inn, Ashover is featured in the Central Television series *Heart of The Country.*

Brian Oldham retires after 34 years, Derek Tarr after 31 years and John Ruskin after 22 years.

More pubs are added to the estate.

Upwards of 12% of Mansfield Beers is being exported to the Baltic States.

1995 Embark on the Investor In People accreditation programme.

Invest £3.5 million in a new distribution centre to be operational from October 1st.

March: The Office of Fair Trading begin a three month enquiry into wholesale beer prices. Mansfield Brewery was one of the first breweries selected to make a submission to the OFT enquiry.
The message to the government is clear; 'You've got it all wrong.'

July: Profits for the year ending April 1995 are up by an impressive 21.5% to £17.4 million. The 1200 qualifying staff are celebrating a £¾ million profit share.

August: Darrell Stocks joins as Retail Director.

September: 30 year old Richard Aldread is awarded a B.Sc. in Building having completed a four year part-time course at Nottingham Trent University. Richard joined the company at 16 years as an apprentice bricklayer.

During the year a new initiative is encouraging country locals to become the hub of their local community by introducing small 'shops' into their pubs.

1996 £2.4 million investment in brewing equipment. By late summer Mansfield Brewery becomes the 7th largest brewer in the country.

January: Gordon Kendall retires after 50 years. He says he can remember getting his first week's wages of 26 shillings (£1.30) and 'feeling very rich'.

February: Contract brewing of all Home Brewery products is obtained from Scottish & Newcastle Breweries, following the closure of Daybrook.

July: A new £15,000 Chairman's Award Fund is launched. Individuals or pubs working within their communities can turn to the fund for help.

The Labour Party's Deputy Leader John Prescott and Parliamentary Beer Club Chairman Alan Meale joined Commercial Director Colin Stump in a protest against beer tax.

August: Jeremy Clarkson is the latest star of the Mansfield Bitter TV ads.

Harry 'Nobby' Beighton retires after 41 years.

October: Mavis Leverton retires after 23 years.

Mansfield voted the East Midlands' favourite brew by Daily Mirror readers.

December: On the 26th Sales Director Dennis Foster dies peacefully at home. It was a great shock to everyone. He is greatly missed throughout the Trade.

1997 February: The sole surviving heir of the Baily family, Frank Bruce Baily Thomas, dies.

June: Eight employees spent two days in the Lake District climbing 24 peaks for charity. It is anticipated fundraising will reach £15,000 for Childrens Aid Direct.

Richard Chadburn relinquishes his executive directorship, but remains on the board as non-executive.

July: Profits for the year ending 29th March are up 13% to £20.8 million.

John O'Neill joins the company to head up Free Trade sales.

August: Jack Ward retires after 33 years. His wife worked at the William IV for more than 12 years.

October: Redeye, a unique premium smooth beer, is launched.

During the year the new pub concept company is launched. Concepts to date are Trading Post, Lloyds No.1, Lock, Stock & Barrell, Brownes and Xtra.

7 pubs were purchased in Northampton from McManus Taverns.

1998 January: Carlton Television asked for help for a scene showing a beer delivery in their hit series Peak Practice. The episode featuring driver Chris Warner and his mate Herbert 'Teletubby' Roberts was screened this month.

Cellar Services engineer Dave Briscoe has been sworn in as a Justice of The Peace in Hull.

March: Michael Ruddell, Personnel Director of the Boots Company, joins the Board as a non-executive Director.

April: Ken Smith, Harry Fisher and Eric Murray retire after working at the Brewery for 99 years between them.

June: Ken Johnson, the Poacher, Rossington, died on the 8th and Terry Widdall, Oddfellows, Beverley, on the 5th.

This year's Lakeland Challenge raised £7,000 for charity.

July: The *Making It!* Discovery Centre opens, Managed by Sue Walters. She left her position of secretary/PA to the Managing Director to develop the discovery centre concept.

August: Ken Hurt retires after 34 years.

December: Hugh Wilson, Property Director, retires.

During the year a two year £6 million investment to refurbish 50 outlets commenced.

Friar Tucks and T. P. Woods added to the list of concept pubs.

Bill McCosh awarded the CBE for his services to training.

1999 Downturn in trade causes a transfer of 60 managed houses to trade partnerships involving 98 managers and 650 staff. Also more than 20 redundancies take place in Mansfield and Hull. The financial crisis in the east means the Brewery loses its valuable Russian contracts.

June: The Brewery is awarded the coveted *Investor In People* award.

August: Sir David White retires after the Annual General Meeting.

Millennium Ale in a presentation tin for supermarket sale is produced.

December: All employees receive a presentation tin containing a bottle of Millenium Ale, a badged glass and a bottle opener.
During the year Mansfield Dark Smooth Ale appears.

Almost a century and a half of Mansfield's brewing history draws to a close as, on December 21st, Mansfield Brewery is sold to Wolverhampton & Dudley Breweries.

✠ ✠ ✠ ✠

Edward Deakin, Joint Managing Director until retirement in 1971

World War 1: Edward Titley, son of Addison Titley, in military uniform.

Graham Wilson, who worked in Cellar Installations Department, retired after 44 years in 1979 but died shortly afterwards.

Alan William Chadburn, son of Claude, killed at Tobruk June 6 1942.

Early 2000: Mike Cleator: One of many snaps taken by Vicky Phillips as the sun set on the Brewery.

Against All Odds : British Beer – A Success Story.

'Designing a beer to sit in a particular place is quite interesting. You approach it from a standpoint of right, well, I want it to be this colour, this level of bitterness and that's the sort of flavour that I want to achieve, and you select all your ingredients to achieve the flavour that you want. There's a lot of experience in it, yet there's still a lot of witchcraft in brewing."

Mike Cleator, Chief Engineer
Brewhouse and Fermentation Manager,
Mansfield Brewery.

The above words are those of a master magician. Mike Cleator's allusion to witchcraft may reveal his innate sense of humour, (something you'll find common among brewery workers) yet he's right. One could hardly attribute such care, passion and witchcraft to some of today's chemically-formulated mass market beers. A good brewer succeeds or fails by his attention to taste and consistency. Drinkers may not realise that 'witchcraft' is involved, but if the 'magic' goes out of their ale, then they'll soon let you know.

Along with his colleagues and his predecessors, as a brewer Mike Cleator was the guardian of the secrets of success. This book deals with the pride in that success, and is mainly a chronicle of those last free, independent decades of existence for a great and much-loved local company, Mansfield Brewery, as told by the men and women who worked there. It is the story of people who loved their work, often having no desire to work anywhere else. There was of course one less favourable rule of employment which has been standard procedure in many industries. Sacking employees who had reached the age of 18 rather than retain them on a man's wage was a common occurrence. If your aptitude and ability looked promising, you may, as many of the participants in this book discovered, be kept on. However, those who remained enjoyed an environment which has vanished since the 1980s – the workplace where employees were actually held in regard by management, where skill, loyalty and length of service meant something.

Towards the end of the 20th century the British Isles had up to 300 breweries. The largest was Bass, with over half a dozen brewing centres in the UK turning out millions of gallons of varied beers and ales under a variety of trade names. Many of those brands had been absorbed into the Bass empire throughout that huge company's long history. Over the last 50 years, the history of brewing in the UK is one of a rapacious quest for monopoly. Beer is big business. As drinkers, we may think of it more romantically as 'the staff of life' but to organisations such as Bass and Whitbread it is the stuff of profit. Like fine wines, beers have their own subtle signatures, regional flavours and followings, and have always contributed greatly in making British pub culture stand out from that of the rest of the world.
Surviving the constant hunt of the late 20th century corporate wolf pack was no mean feat. The fact that Mansfield's Brewery maintained its sturdy independence for well over a century and a half is something the town can be proud of. Many world famous international brands, which, in the 1950s, seemed like national institutions, indestructible and untouchable, would eventually be caught up in the big brewers' net. No one in Europe four decades ago could ever imagine that age-old beers such as Carlsberg, Stella Artois or Grolsch could ever end up simply as brand names, brewed on licence in Britain, their taste and consistency often bearing scant relationship to their European origination. Each new 'cult' drink, imported on the back of heavy marketing campaigns, such as Australia's Fosters or Castlemaine XXXX, Jamaica's Red Stripe and others, could never survive the financial blandishments of the brand-collectors. It seemed for a time that individuality in British beer might well be stamped out – yet there have always been brewers in the UK who would hang on fiercely to

their independence, with their individual creativity and local support a continuing pillar of strength. Mansfield Brewery was such an organisation.

Throughout Britain's history, Mansfield in Nottinghamshire has been known at various times for various reasons. In the Middle Ages it was the centre of a lush, natural forest – Sherwood, the hunting domain of Kings, mythically the battle ground upon which Robin Hood ran his guerrilla campaign against an unjust monarchy and its robber Barons, (although many Yorkshiremen will dispute this claim). In later times the rich seams of coal beneath the soil of Nottinghamshire brought to Mansfield a new reputation and prosperity. Within two short decades, from 1984 into the new century, all but one of the many coal pits, once providing work for thousands of thirsty men, have closed. Losing a major industry is one economic tragedy; but when a town like Mansfield loses its Brewery, despite a noble campaign to keep it running, the feeling of insult added to economic injury runs deep.
From its founding in 1855 through to its take-over by Wolverhampton & Dudley Breweries plc in 1999, with the loss of over 200 jobs, Mansfield Brewery has gained an enviable reputation for quality beers, and has often led the way with the innovative introduction of new brands. This was the Brewery which, in the immediate post-Perestroika period in the Soviet bloc, through adventurous management, astute salesmanship and a keen eye for developing new markets, soon found itself exporting bottled beer to Russia worth over £5 million in profits to the Brewery.

Sydney Deakin mashing, Mid 1930s with Tom Hinchley assisting

This book seeks to encapsulate the varied memories of the men who made and sold that beer – not only to the appreciative Russians, but to millions of discerning beer drinkers throughout the UK.

British beer is one of this country's great success stories. The variety and quality of the beers available throughout the UK is also a triumph for beer drinkers, and, in many ways, a mark of success for the Campaign for Real Ale. As the monopolies in the brewing industry continue into a new century to absorb one another in a ravenous quest for even larger shares of this vast market, we can look back to a time in the 1970s when the very character of our ales came under serious threat. Led by that universal brew, Watney's Red Barrel, the wide range of choice we had always enjoyed appeared to narrow rapidly. But there is nothing – in Europe or the world – to match the British pub, and those who challenge the taste of its customers do so at their peril. Public demand insisted that we hung on to our brewing traditions, and good, independent breweries such as Mansfield's led the way in satisfying the need for choice, quality and variety.
Unless you are a fully paid-up member of CAMRA you might not think too deeply about how your beer is brewed or who brews it. A good pub and a good pint go together – you find your local and you stick with it. But brewery workers and those special people – the Brewers – are a breed apart. As you will discover in these pages, the old adage about 'couldn't organise a booze-up in a brewery' has a special resonance here. These men could organise one anywhere. There is a special attraction to brewing beer as a job. Many of Mansfield's brewery workers were in service for the greater part of their working lives – some as long as 50 years. From the sales office right down to the bottling plant the social division of 'us and them' which dominates many other industries had a less prominent place in brewing in later decades. Making good beer has been a pleasure for these people, and the camaraderie they've enjoyed over the years is obvious.

Time and the tide of capitalism wait for no man, and the eventual take-over of Mansfield Brewery by Wolverhampton & Dudley in 1999 was a sad event for the people of Nottinghamshire. But the famous brew did not leave the town without a spirited fight. A strong campaign fought hard to keep the Mansfield brew on its home turf, but modern commerce pays little heed to nostalgia or fond emotions. The combined forces of banking and accountancy are, as ever, the dictators of our business history, but at least we have the legacy of the Mansfield trade mark, along with a beer which, although no longer brewed in the town, remains as popular as ever in the North and the Midlands.

This book came about through the determination of Mansfield Brewery's Head of Personnel, John Else. John, along with other ex-employees, saw the need for the memories of his work mates to be permanently recorded. So inevitably this is a tribute to the men and women who built Mansfield Brewery into the coveted corporate gem it became. It is not exactly a nuts and bolts history of brewing technology, or a compendium of facts, figures and dates. This is a brewer's first-hand history, a collective memoir of the men who moved the hops, who shovelled the barley, who ran the water, as well as those who wore the suits and the women who sat behind the administrative desks. It is a pleasant memory of an industry devoted to making us feel good. Hopefully, we can all drink to that.

❖ ❖ ❖ ❖

Chapter One
Birth of The Brew

'We shall be true and just to each other'.

Words from the original Mansfield Brewery Partnership Deeds between John Watson, William Baily and Samuel Hage, 1855.

It began in Littleworth, a compressed thoroughfare down by Mansfield's narrow River Maun – more of a glorified stream than a raging torrent. And there it remained until the last brew was made a century and a half later.

Samuel Hage was a farmer out at Ollerton. John Watson was a brewer in Sheffield. Records show William Baily's occupation as 'Gentleman'.

These were the three men about to embark upon an enterprise which would, over the next 145 years, provide millions of Britons – and, later, overseas drinkers – with the pleasure gained in drinking good ale.

The mid 19th century was a great time of expansion and industrial innovation. The railway had arrived, and although there were still less than 2,000 miles of track throughout the United Kingdom, they were being rapidly developed. The area around Mansfield already possessed a fine reputation for its profitable agricultural prowess, but it was deep down beneath the soil where the new industrial gold was being exploited. Coal was definitely King, and the miners of Mansfield were very thirsty men.

Up to the end of the 17th century the tradition of pubs brewing their own ale on the premises had reigned supreme. However, as brewing on a larger, wholesale scale grew into a specialised craft, the pecuniary benefits of satisfying a growing industrial workforce's thirst began to look like an attractive investment.

The decades from the mid 19th century to World War I represented a whirlwind of commercial activity for the Brewery. Pubs were bought up to become Mansfield houses at a remarkable rate, and with the coal fields in full production, business was brisk. There were various exits and entrances on the board of directors over those years, but the Brewery did not become a Limited Company until as late as 1925.

One of Mansfield Brewery's later memorable helmsmen, the redoubtable William Jackson Chadburn, passed away in May 1922. Until then the business had been run as a 19th century partnership, and to bring the Brewery into the modern commercial age, the lengthy procedure to dissolve the partnership began. It took almost three years, but on February 7th 1925 the company became Mansfield Brewery Limited, with William Jackson Chadburn's son, Claude, as Managing Director. But behind all this and subsequent activity in the boardroom there is a hidden story, at times rather touching and unusual, the outcome of which, in many ways, spurred the company on to even greater success. This is a delve back into the Brewery archives, yet it is a worthwhile detour. It is a mini-saga only John Galsworthy could emulate; in turn tragic, dark and poignant, yet woven into the true fabric of a great enterprise.

A Blameless Life: Frank Bruce Baily Thomas 1916-1997

William Baily, one of the original joint founders of the Brewery in 1855, had a son, William Edward Baily. In turn young William Edward married and had a daughter, Gertrude Sarah. She married one F. K. Munro Thomas in the early days of World War I. Thomas had worked in Worthing as a Junior Master at St. Dunstan's Preparatory School for Boys. Soon after marrying Gertrude, with the war dominating Europe, he was commissioned into the army, eventually achieving the rank of Captain. It was a title he obviously enjoyed, as throughout his working life, everyone referred to Munro Thomas as 'The Captain'. We cast no aspersions here, but, in addition to the joys of courtship and marriage, Gertrude must have been quite an attraction to Munro Thomas who seemed, in his choice of a spouse, lucky in more than just love. When her father, William Edward Baily, passed away, his substantial share – 50% of Mansfield Brewery Ltd., was left to his widow to be divided equally between her and Gertrude under a Trust created by his Will. Upon her death, that 50% share and profits would pass to Gertrude. During her widowhood Mrs. Baily enjoyed the increasing dividends from the profits of this. Early in the war Gertrude received a letter from the Brewery partner, William Jackson Chadburn, informing her of how well the business was doing. War or no war, they were selling plenty of beer, as in addition to the regular coal fields consumers, an Army camp had been set up at Clipstone, housing 60,000 thirsty squaddies.

There is an odd side-note to Gertrude's relationship to William Jackson Chadburn. He had become her trustee upon the death of her father, and the records seem to show a close relationship in all matters surrounding the Brewery. Yet Gertrude never told her Trustee that she had married, and always accepted her letters in the south of England under her maiden name as 'Miss Gertrude Baily'. Chadburn did not discover the truth until Munro Thomas, then a Lieutenant, came home on leave in 1915 after being wounded. It is difficult today to ascertain what lay behind this subterfuge, but there is a hint that Gertrude's new, and, for a while, unknown, husband had also become her financial adviser. When Chadburn had realised that Gertrude's share of the profits in that first year of her marriage would be the then princely sum of £6,000, he had asked if she was prepared to leave £3,000 in the company's coffers. This would be seen as a gesture of support in the company's attempt to reduce its indebtedness to the bank. Gertrude, however, said that she was going to use all her money and convert it to War Loans. Although this sounds very patriotic, and no doubt, being married to an officer, she would have been as jingoistic as any Army wife, one cannot help but suspect the hand of 'The Captain' behind this. Such hefty sums of cash passing through the bank account of the wife of a soldier, quite recently working as a teacher, must have been a heady prospect for the keen new husband.

Frank Munro-Thomas

In 1916 Gertrude and Munro Thomas had a child, Frank Bruce Baily Thomas. Gertrude by this time was 32. Shortly after the Great War ended, the unfortunate couple came to realise that their young son was mentally handicapped. It was undoubtedly a severe blow to the couple. A further tragedy occurred only seven years after Frank's birth, when, in 1924, Gertrude Sarah died at the early age of 39. One might have expected that Captain Thomas's tragic loss of his wife may have been lightened by the passing of her substantial Brewery fortune to him. But this was not to be the case. However, from her own capital, and not her Brewery money, Gertrude had left her husband the fruits of a Marriage Settlement which had decreed that Munro Thomas would at least receive an income. That said, it was not to be as spectacular as the one he might have received if Gertrude had left him the Brewery shares. He would not be affluent, although throughout his life he would be 'comfortably' provided for. It was to be poor little Frank, who would not achieve a mental age above 9 years throughout his life, who would finally receive the half share in the Brewery when he came of age at 21 in 1937. When The Captain finally passed away in 1970, Frank was also bequeathed the capital of his mother's marriage settlement. Thus Frank Bruce Baily Thomas, a child-like innocent on the fringes of the big, rapacious world of commerce, was to become a very important young man with money to match.

Upon Gertrude's death her closest friend, Miss Molly White, had been appointed as young Frank's guardian. Whilst Frank was still a minor, between 1924 to 1937, Molly shared the house with the Captain and her young charge. As a widower, and still a relatively young man, sharing your home with a young woman, officially your son's guardian, would surely result in an eventual emotional attachment. And Frank's legal status as a man of some wealth after 1937 would inevitably create some enmity and jealousy in later years. The Captain may have thought that he was getting a little closer to the fortune when, in 1937, he was appointed as young Frank's Receiver. The law stated, under the Court of Protection rules (which were known in those days as the Management and Administration Department of The Court), that due to Frank's mental age his inheritance should be legally controlled. Munro Thomas, however, did not find the sometimes thankless task of being his son's Receiver to his liking. He simply couldn't get his own way, and after six months of bitter argument with all concerned, he resigned the job early in 1938. A new man, Mr. Michael Bennett, of the accountancy firm Barton, Mayhew & Co., filled in until Mr. Bennett went off to war in September 1939. But Bennett seems to have been an industrious man. On Frank's behalf he bought an impressive period home, complete with a 100 acre farm, just outside Horsham. The unusual trio of Molly, the Captain and Frank moved in, together with their own staff to run the house, grounds and farm. They had taken with them Cecil Binstead, who had been on their household staff at Worthing, and now had a chauffeur and handyman called Ted Slaughter and his wife, Peggy, the cook. Molly White seems to have done well during this period. She bought herself a second house so that she could occasionally stay in Worthing, and this residence was cared for in her absence by Cecil Binstead's wife, Winifred. With the war gathering momentum however, Cecil Binstead was soon to enter the RAF. With Michael Bennett now in the forces, a new Receiver for Frank enters the story.

In 1939, Jocelyn (known as 'Jock') Nangle, had been ineligible for military service. During his childhood he had suffered from poliomyelitis. As Frank's new Receiver, Jock was a frequent guest at the Thomas estate in Horsham, but he didn't find his stays there to be very comfortable. The Captain's frustration and bitterness had spilled over into some kind of adverse domestic relationship with the cook, Peggy Slaughter. Looking back, it is hard to hazard a guess at the psychological strains in this household from this vantage point in time, but a little imagination can provide the scenario.

Munro Thomas marries a potentially rich young woman. He serves his country, is wounded. They have a son, a happy event which is soon to be sullied by the tragic news of his mental impairment. After only eight years of marriage, Gertrude dies. Although the Captain is provided for under the marriage settlement, his existence will be somewhat removed from the one he may have expected had Gertrude left him her Brewery portfolio. Then along comes Molly White in the official role as little Frank's guardian. The years pass; as Frank reaches maturity, his father has to watch his mentally impaired innocent offspring become exceedingly well-off. The Captain then discovers that even as Frank's Receiver, he still has little control over events as he is forever under the scrutiny of the legal profession. The atmosphere, therefore, in the house at Horsham, must have been like a pressure cooker about to explode. As the domestic feuding accelerated, and the mental strain of all this began to effect Molly, she appeared more and more to rely on the domestic staff to care for Frank. Jock Nangle was an astute social observer. He had no choice but to suggest to the court in 1941 that, to avoid an even worse situation developing, Munro Thomas had to be taken out of the picture. The court wisely agreed, and Frank's funds provided for a small house to be purchased at Angmering. It was to this new home that the Captain, plus Ted and Peggy Slaughter, moved in 1941.

The bitterness of this period is obvious from the fact that, once he had moved, Munro Thomas expressed the wish that he would never again see his son. It was a sad promise he kept right up until his death in 1970.

One might imagine that the Captain's exit would have made things easier at Horsham, but this was sadly not the case. The enmity and anguish Molly had gone through had left her emotionally drained. Frank may have been a childish innocent, yet he too had been the brunt and indeed the central focus, of most of the disturbances, and this had taken its toll, the effect being some rather outlandish behaviour. Despite his mental infancy, Frank had grown into a young man with all the sexual curiosity one might expect. When Frank was accompanied on shopping trips into Horsham, he would on occasions escape Molly's or the servants' supervision and vanish into the town, where he would demonstrate his growing interest in all things sexual by approaching young children. Thus the beleaguered Jock Nangle on more than one occasion had to face the wrath of Horsham's Police force and talk young Frank out of trouble.

Frank needed a male 'carer' so Mr. Nangle appointed a retired Naval Officer named Withrington. Based in the Lodge on the estate, Withrington became Frank's shadow and companion, and managed to steer his interests away from any potential sexual misbehaviour. Withrington was eventually replaced by the returning servant, Cecil Binstead, who had now been de-mobbed from the RAF.

When Frank was 35, in 1952, Molly White died. After buying a bungalow on the South Downs at Durrington, the Binsteads volunteered to look after Frank. The Horsham estate was sold, and soon the Baily Thomas residence would be replaced as a large new house was bought for Frank. This had a spectacular four acres of grounds, and was peacefully situated next to Worthing golf course. The Binsteads sold up and moved in with Frank in the summer of 1952.

After all the acrimony and rigour in his previous years within in the strained atmosphere of the old Horsham estate, Frank was at last able to enjoy a peaceful and stress-free life. This was a result of the genuine, loving care of the Binsteads over three happy decades.

Yet the rigours of age show no mercy. By the 1980s the Binsteads had entered their 70s. Maintaining the large house and grounds was taking its toll and their health was failing. To add to this state of affairs Frank had suffered a serious accident and had broken his hip. The big, modern house which had been such a boon to the trio in earlier years seemed to offer an insurmountable challenge. Frank's mobility was now restricted to a mix of crutches, a wheelchair and walking frame. Getting the poor man upstairs was a daily trial.

Once again, the reliable Jock Nangle came to the rescue. After heartfelt discussions it was decided that a new bungalow should be built at the end of the house's substantial grounds. This new residence was aptly named 'Franklands' and was completed in 1982, when the big house was duly sold, along with two acres of the grounds for building development. Once again, a new and happier period had dawned for Frank and the Binsteads. At last he could move around freely, with plenty of space for his wheelchair, and the Binsteads could finally shed the burden of caring for the big house and the extensive grounds. An examination by psychiatrists at this point resulted in an assessment that Frank's mental age was now somewhere between 8 and 9 years.

Winifred Binstead passed away in 1984, and Cecil died from a heart attack sustained whilst driving his car in 1986. At his death, the hapless Frank was Cecil's passenger. Although he bore the loss with some bravery, the two deaths must have marked one of the lowest points in Frank's challenging life. Cecil and Winifred had become like a mother and father to him, and their circle of relatives formed what can only be described as his real family. The day of Cecil's death was also a piece of bad timing for the over-burdened Jock Nangle. As Cecil breathed his last, Jock was 30,000 feet over the Caribbean with his wife, ready to land for a much-needed holiday. But there was to be no joyous unpacking of his luggage. He was on the next plane back to England where he headed straight for Worthing and opened discussions with the remaining members of the Binstead family. The kindness which seems to have characterised the Binsteads was amply reflected in the offer of the departed Cecil's niece and nephew, Marian and Richard Wren. Both gave up their jobs and moved into Franklands, thereby enabling the shocked and sorrowful Frank to pick up the pieces of his shattered existence with people he already knew and loved. And so he remained, the childish innocent link between the growing

Mansfield Brewery and 50% of the company's shares. It may be a more rounded story had Frank lived out the life of the Brewery itself until 1999 – but sadly, at the ripe old age of 81, he died peacefully in 1997. One cannot but wonder what the course of events may have been for the Brewery had fate not dealt poor Frank the hammer-blow of his mental condition.

Frank Bruce Baily Thomas's life is the emotional, more human facet of the great story of Mansfield Brewery Ltd. His death opened up a more commercial chapter; who would get hold of that crucial 50%?

A Blameless Life: Part 2:
Jock Gets Tough.

There may be some vague camaraderie in company boardrooms, but beneath the millpond surface of the corporate ocean everyone's a shark. More minor financial dealings of a risky nature can sometimes be conveniently buried in the pages of the annual report & accounts, often to be missed by even the most eagle-eyed shareholder. But subjected to the doctrinaire and thorough inspection of Jock Nangle, one such deal proved to be a major turning-point. When Jock Nangle was appointed Frank's Receiver in 1939, Frank owned approximately 51% of the Brewery's Ordinary Shares. Despite the war's high taxation rates, dividends were still adequate enough to keep Frank comfortably in his Horsham residence with his retinue of staff. Once he'd moved in with the Binsteads in Worthing, the financial situation was even better.

The Chadburn family were still managing the brewery as substantial minority shareholders – and managing it very well. Safe in the knowledge that Frank's 51% interest had remained intact, Jock Nangle lived by that old adage, "if it ain't broke – don't fix it." Each year he combed through the company's accounts and, although not spectacular, what he saw was steady upward progress.

The accounts for the year ending March 31st 1956 seemed to demonstrate yet another good year – but Jock noticed something

Claude W. Chadburn

amiss. That year Mansfield Brewery Ltd. had purchased a small wine and soft drinks merchants. Nangle noticed that the price for this transaction had been satisfied partly in cash – topped up with Ordinary Shares in the Brewery. This had resulted in Frank's holdings being reduced from the previous 51% to 48% - without consultation to Frank's Receiver.

Jock was understandably not best pleased with this situation, and let his concerns be known in no uncertain terms. He referred the matter to the Court of Protection and then fired off a broadside to the Chairman of the Brewery, Claude W. Chadburn. Significantly, in his letter to Chadburn, Jock Nangle demanded that he should have a seat on the Board of Directors. As a man holding 48% of the company's equity, the Brewery had no other choice, and Claude Chadburn immediately agreed that Jock should join the Board. With both of Claude Chadburn's sons having died young, he sought to 'balance' the arrival of a new director by appointing one of his own; in this case, his nephew, Robin W. Chadburn.

The arrival of Messrs. Nangle and Chadburn jnr. in the boardroom was to prove one of the most fortuitous events in the history of Mansfield Brewery. It was the harbinger of an unprecedented period of expansion and rising profits, which would, somewhat perversely, eventually turn the company into the very juicy bait the venture capitalists of the 21st century were looking for.

Yet at the time Jock Nangle joined the Board, his first priority was the thorny question of what was to eventually become of Frank's portfolio.

With Frank's death, and with him having no living relatives, his Brewery shares could well fall into the grasp of the Crown as they could be liable for 80% death duties. With half the company's stock thus vanished, Mansfield Brewery could easily lose the independence

which it had enjoyed for so many years.
The story of how Jock Nangle prevented this from happening is too complicated and involved to outline here in full, but it is a testament to his sharp thinking and the brilliant utilisation of his profession as an accountant. Suffice it to say that the end result was, that through diligence and patience, Jock managed to gently cajole the Chancery Division, following Frank's death, into approving a scheme which prevented Frank's £2½ million fortune from vanishing. A laudable Charitable Foundation was established whose objectives were to carry out research into mental illness and to promote the care of those who suffered from this disability. It seems a fitting and poignant reminder of a simple, yet very important man, Frank Baily Thomas.

Jock Nangle left his career as an accountant in 1976 but continued his work with both the Charitable Fund, and his other project, the Provident Fund, which had been set up to benefit the employees of Mansfield Brewery, as well as devoting much of his time to the affairs of the company, where he became a good friend and colleague to his co-director Robin Chadburn.
During his years prior to leaving accountancy, he had often commuted between his job in the City of London with Barton Mayhew & Co. and Mansfield Brewery. This formidable and compassionate man of figures no doubt contributed as much to his City post as he did elsewhere, as Barton and Mayhew, through a route of take-overs and mergers, finally became absorbed by the giant Ernst & Whinney, one of the UK's 'big eight' accountancy firms, now known as Ernst & Young.

❖ ❖ ❖ ❖

Draymen, 1930: Left to right: A. Boxford, T. Guilor, F. Holmes (horse trader) S. Dent, F. Groves, H. Boddy, R. Gilliatt, J. Marriott

Chapter Two
A Lot of Bottle

'They kept us in order – but we got the job done...'
John Bartle, Wines & Spirits 1974-1989.

'My first job was 'rousing' rum...I got drunk just on the fumes...'
Sam Sills, Wines & Spirits 1962-1999.

To hear men talking about their career at Mansfield Brewery now, only a few short years after its demise, brings to mind a kind of working life which has gone forever. Yet what a life it was – making beer, bottling wines and spirits, running pubs, buying the best beverages from around Europe. Today, in the company of MBC's one-time Head of Personnel, John Else, and Wines & Spirits Manager John Bartle, who still continues in the trade with the new owners of the Mansfield brand, Wolverhampton & Dudley Breweries, we set out on a mission to meet one of the surviving oracles of those great times down at Littleworth. Like John Bartle, Sam Sills has spent most of his working life in the wines and spirits trade for Mansfield Brewery Plc. and IDV (International Distillers & Vintners). Today he runs a quiet country pub in Leicestershire, noted for fine food and, as one would expect, good traditional ale.

1964: A cut above: John Bartle on the Brewery's office roof, sawing up unwanted stillages for firewood.

We arrive late on a damp February afternoon, and in the cosy bar the beer soon begins to flow along with Sam's memories.

"I was 17 when I started at the brewery in 1974. I'm a Sheffield Wednesday fan and they were playing Aston Villa in a replay at Villa Park. I decided to take an afternoon off to watch this. This was when we had the old second and first divisions. I'd come home from work this dinnertime – I was a trainee manager with the Co-op, where I'd been for about seven months.

Anyway, I rang in to work and put my dad's voice on saying 'He won't be in this afternoon – he's took ever so bad…' Well – just my luck – Co-op's area manager saw me on the market place wearing my Sheffield Wednesday scarf and hat, getting on the bus. He collared me the next day about it but I said 'But we *beat* Villa….' and he said 'What's more important to you – Sheffield Wednesday or your career?' I said 'Well, let me put it this way – now we're into the 5th round – Sheffield Wednesday!' He sacked me – but in any case, actually, I resigned. Well…alright then, I was sacked…."

I got on a bus to Mansfield. I had about £8 or £10 in my pocket and I'd seen a shirt I wanted, so I thought I'd go through to Mansfield and get this shirt. When I got off the bus I met this bloke – Ron Evans. He was wearing a Mansfield Brewery badge and I asked him where the brewery was. He told me, so I thought I'd go down there and see if there were any jobs going. I still had my father to face with the news I'd been sacked.

I remember going through the brewery's revolving doors and there was the receptionist, Madge, sitting there. I said 'Is there any jobs going, Missus?' and she replied 'Oh…have you come for that job in the wines and spirits department?' So I said I had.

So she asked me to sit down and wait and eventually Harry Ridley, who was the wines & spirits gaffer, came through to fetch me. When we got into his department I recognised one of the lads in there -–I knew him. Anyway, to my surprise Ridley set me on and said 'When can you start?' I said 'Tomorrow.'

So the next day I started work with the brewery. As I was walking out after being given the job, there was this lad sat in reception, all dressed in his best suit with the biggest knot you could imagine in his tie. Ridley said to him 'What do you want, son?' and the lad said 'I've come for that job in the wines and spirits department..' Ridley said 'Oh, that's gone, that has…'

Actually, I'd got that poor lad's interview – and his job!

And that's how my career started in the drinks trade – as John Bartle's tea boy....no, actually, as a warehouseman. We imported all our own wines and sherries and I was involved in bottling them. Spirits came from all over – the Caribbean, Scotland, wines from France... they were great days. I was 17 and at that time I'd never taken a drink. I'll never forget going back to my dad that day though. My dad was a bookie. I got home later that day and he was in the bathroom having a shave, getting ready to go to the dog track that night. I said 'Oh... by the way – I got sacked today.' He swung around with this cut-throat razor in his hand... for a minute I thought he was going to cut my throat – and he said 'You little bugger...' I told him what had happened with the area manager at the co-op, and said that I'd told him to shove the job. I said 'But – before you hit me, I've got another job – at Mansfield Brewery.' He paused for a moment, then looked at me quizzically, then asked 'Do you get free beer?'
I said 'Yes, I do...two bottles a day.'
So he smiled and said 'Good lad!' And thankfully he left me alone."

"My first job at the brewery was 'rousing' rum. When it came in in bulk from the Caribbean it was 170% proof. It was then put into glass-lined tanks and we added distilled water until it comes down to 72 or 73%. Then somebody has to hang over the top of the tank with a stick with chains on – that was me – and keep stirring it. You had to get right over the tank and stir it for about three hours. God, I was badly after that. My dad thrashed me and said 'You're pissed up!' I told him I hadn't touched a drop and it was true – it was the fumes from the rum. I was staggering all over the place, I threw up on the carpet. Dad thrashed me to bed and I've never touched dark rum since. That was my introduction to Mansfield Brewery.
I was lucky because we had a gaffer there, Les Pearson, who was a very interesting man. More importantly he was a man who believed in teaching you things. This meant that they weren't frightened to open a bottle of port and letting you try it – after all, it was part of your training – they'd say 'Come and try this.' Yes, it was a cracking company to

work for in those days…"
At this stage of Sam's career, John Bartle was already a veteran in the wines and spirits game. John also recalls Les Pearson.

"Les was an expert in vintage port. He could taste any unmarked bottle and accurately tell you the year. When Les retired – about 1979 – that morning he said 'I think we'll have a drink of port.' He told me to bring two special bottles up from the cellar, which we decanted over the sink in the toilets. It was very old port – when decanted it had lost a lot of its colour – it looked like a good rosé wine. Les sent one of the lads out for some cheese from Longford Dairies in Mansfield – really good Stilton. When he poured the first bottle, the capsule or cork would usually show the date. In this case it didn't. Les said he reckoned it to be 1897 or 1908. We had four large glasses poured out – but when we went back half an hour later the air had got to the port and it now tasted like vinegar – but those first glasses were superb. Anyway, there was a date on the second cork – and sure enough, Les was spot on – it was 1908! In fact, up until 1980, we sampled and drank some of the finest wines there were – especially port. We were working class lads on £15 per week and we drank every best vintage port of the 20th century. These were wines which would usually only be tasted by the very rich. In those days all the finest wines only went to two locations from Mansfield – the White Post and Chesterfield's Station Hotel. The reps used to bring samples in. In those days the company had no worries about laying down £2,000 worth of port because within six or seven years it was worth over £5,000. But we reached a stage in the 1980s where accountants took over and they'd rather have the money in the bank than laid down in good wines."
Good wines, such as Nuit St. George or Volnay were stock-in-trade for Mansfield Brewery – and whenever a customer opened a bottle they could guarantee quality. However, as John recalls, what was on the label wasn't always what was in the bottle.
"In those days we only had two major wine outlets which were the White Post and Chesterfield's Station Hotel. One of the directors, Hector Chalmers, ran the Station Hotel and I can remember Harry

Ridley of the wines & spirits department coming in to us one day and saying that they were concerned at the Station because we'd crossed Nuit St.George off the list. I said 'Well – we've got none, Harry – its all gone...' to which he replied 'Well – we've got some Volnay, haven't we?' I said we had; 'Well, stick Nuit St. George labels on it – they'll not know the difference...'

The young John Bartle

"Looking back, with all due respect to Nuit St. George – they were getting a better wine with the Volnay – these aren't wines you could afford today – they'd be up to £40 a bottle".

It was a rare occurrence indeed but sometimes bottles contained other surprises, as Sam Sills pointed out;

"When I was a lad in wines and spirits I remember I'd had cereal for breakfast one morning. This would have been in the mid 1970s. Kelloggs at the time used to put these little clip-together battleships in cornflake packets and I sat there on the bus one morning putting this thing together. I looked at it, thought 'that's smart' and stuck it in my pocket. I forgot all about it. A few hours later that day we were bottling sherry and I felt in my pocket and thought 'what's that?' - it was the battleship. Well, it was clogging my pocket up and I got rid of it – just chucked it away. Six months later Les Pearson gathered us all upstairs and said 'We've had a serious complaint from Mrs. Chalmers at the Station in Chesterfield...she was pouring out a fine Amontillado last night at a customer's table and apart from sherry she got a plastic battleship in her glass...'

Probably today such an event would signify a marketing campaign where the lucky consumer wins a trip to Gibraltar – or perhaps that particular Amontillado was the 'Portsmouth vintage.'

Sam recalls some of the odd duties staff were sometimes called upon to perform.

"I think it was late 1975 or early 1976. I recall our receptionist, Madge,

1973: Message in a bottle? John Bartle dusts down his knowledge of vintage port with this 1963 Croft – the last to be bottled by Mansfield Brewery in 1966. (We wonder - what became of it?)

asking 'Can somebody send a lad up to Mr. Scruby's office?' Scruby was then the Managing Director. I was selected and told to take a dustpan and brush with me. He used to take his two gun dogs to work with him. They used to lie under this big, old-fashioned radiator and, as you'd expect, occasionally they'd do their business on the floor. When I arrived in his office, I'd imagined that he'd broken a glass or something. But no – there was dog crap all over the carpet. Well, you couldn't use the brush on it – it would just dig into the carpet, so I had to somehow scoop the poop into the pan and hope for the best. Well, I did this job for about three days. I said to Madge 'Why does he bring those dogs into work? It seems cruel. How about asking him if I can take them for a walk?' There was a park opposite, part of which we eventually built the new Mansfield Brewery offices on. Surprisingly, he agreed to me walking his dogs – and they were lovely animals – in the park. About three weeks later we had a union meeting. I remember there were two union officials who used to run the meetings – a fat bloke and a thin bloke. When points of order were being raised, I stood up. The chairman looked over and said 'Yes – Brother Sills?' I said 'During the last few weeks I've had to go up to Mr. Scruby's office every day to clean dog shit up. I don't think he should bring his dogs into work every day. Anyway, cleaning their muck up isn't very nice.' The fat union bloke whispered to the thin one and said 'Brother Sills – what are you employed as?' I replied
'A labourer.' So he said
'You can't get much more *labouring* than that, lad...."

Tony Scruby had been a farmer in Africa at one time. There was perhaps an old colonial air about him. Sam remembers one of his first encounters with Mr. Scruby, not long after he had taken up the post of MBC's Managing Director.

“I recall one morning we were offloading a delivery of cases of Martell Brandy and Scruby came into the loading bay. He looked across at us and said ‘You there – what are you doing?’
I said ‘We’re unloading this brandy, sir.’ So he walks up to us and says ‘Here – hand me a case.’ So I passed a case to him and he just dropped it on the floor, saying ‘*That’s* how we unload brandy in Africa – now, fetch a glass!’ And so we ended up drinking brandy from this dropped case”.

At times there was a definite flavour of the old Grace Brothers style of higher management as featured in that popular TV show *Are You Being Served.* Sam Sills remembers one occasion when he was labelling wine cases at a busy time of the year when one of the company Elders, Mr. Claude Chadburn, who at the time was well beyond retirement age, entered the room and tapped Sam on the shoulder.
‘You look busy, eh?’ Sam looked up surprised.
‘Yes Mr. Chadburn, we are – we’re packing all this wine – its Easter..’
For a moment the elderly director looked confused.
‘Oh…. Have we had Christmas, then?’

There is an old saying in the forces that ‘bulls**t baffles brains’. One of MBC’s finest bull merchants was the much-loved wines and spirits supremo, the legendary Harry Ridley. He was commonly known as ‘Harry the Wine’ and according to everyone who knew him at the brewery, was one of the nicest and funniest men around. Harry had much in common with his fellow departmental managers – and the general workforce - inasmuch as he liked a drink. Yet under the influence or not, his prowess as a wine buyer rarely let the company down, and on his foraging expeditions around Europe he would often utilise his discerning nose and sharp eye for a bargain to provide the brewery with some classic wines which always paid dividends. Yet there was a pleasantly crafty side to this loveable man, an example of which Sam Sills humorously provides.

To understand such a story we must look back to a different age and abandon today's justified moral tone about drinking and driving. There were times, before our highways and byways became the clogged arteries they are in the 21st century, before the full implementation of the breathalyser, when the liquid business lunch was a common occurrence. For instance, it is a commonly known fact among those in the industry who recall the days of horse-drawn dray deliveries that some draymen could easily down up to 20 pints of the Mansfield brew on a day's rounds. In their case, unlike a motor vehicle, a dray horse always knew the way home – you could call it an early form of satellite navigation. In the higher echelons of brewery employment, the company car-driving area managers and salesmen would also visit numerous hostelries during their working day. That some of them may have conducted their work over one too many tipples is hardly, in retrospect, surprising. But let Sam tell the tale.

"Harry Ridley was a lovely bloke and he liked a drink. However, he was continually being bollocked for bumping into other people's cars when he was parking. However, he'd got around this by telling Tony Scruby 'You know, Mr. Scruby, it really is terrible – these drunks come out of the pub, bump into your car and then clear off without telling you – I'm afraid it's one of the hazards of the job!'

One day there was a wines and spirits trade fair held at the Swan Hotel. Scruby was there, with his big new black car parked in the car park. And so was Harry. But Harry left before Mr. Scruby, and wouldn't you just know it – he pulled out of the car park and scraped Scruby's car all down one side. Harry was a terrible driver. Anyway, the next day he was in the wines and spirits department when he was suddenly summoned upstairs to Scruby's office. He really thought he was for it. A few minutes later he came down again, all beaming smiles. He opened a bottle of Amontillado. Apparently, Scruby had said to him 'Harry – you know something – you're right about those drunks coming out of pubs and damaging your car – I left mine in the Swan car park yesterday and they scraped it all down one side…I can see now the problem you've had all these years!'

Whilst Tony Scruby was telling Harry Ridley this, Harry's gaze was fixed on the window through which he could see the blokes across the road at Redfern's garage – working like mad removing black paint off his bumper!"

John Bartle began his career at Mansfield Brewery as a lad in July 1962. He was there until the very end, and now works for the new owners, Wolverhampton & Dudley.

"My brother worked at the brewery. He was an ex-butcher and he'd come out of the air force. He didn't like butchery too much because in those days you had to kill your own beasts. I left school at 15 with no qualifications, so it looked like a life down the mines for me. But my dad said that I wasn't going down the pit. So I was wandering the streets looking for a job. My brother said that he could get me a job at the brewery just for the summer. At first I said 'No...I'm not going to the brewery…' but I did eventually start there working in the bottling shop. The two men I worked under, Sid Tyler and Percy Holmes – they were legends at the brewery. We worked on sorting bottles. I know we talk about re-cycling today, but back then every bottle – beer, wine, whatever – which left the brewery came back. My job, along with about 30 others, was to stand alongside this conveyor belt when the bottles came along and sort them, putting the right bottles in the right crate. Of course, we always had cut fingers because some of the bottles used to break, we had no gloves and not even any plasters. You couldn't stop, either, because the man in charge of the bottling belt, Snip Bradbury, used to sit at the end of the belt mending pallets all day. If he saw you'd stopped for any reason and you weren't working, he'd soon chuck a pallet block – that's a hefty chunk of wood – at you. It was a really boring job and we used to clock watch all the time. One of the lads who worked with me, was a poor lad who couldn't read nor write, and even if he looked at the clock he couldn't tell the time. Often he'd say to us 'What time is it?' and we'd say 'It's five to twelve.'
He'd grab his bag and start leaving – and it would be nowhere near

twelve, more like half past eight! If you were lucky, when you reached 18 and they didn't sack you (that was to avoid having to pay you a man's wages in those days), then you might progress to bottling beer. It was my brother who landed me a job in the bottling department. There it was a cushy prospect – I was bottling mineral water, but even so, it was a job where you could be out at any time. My brother put the word in for me to move on again, and after he'd spoken to Harry Ridley, Les Pearson interviewed me and after a year bottling mineral water I found myself in the main brewery building in the wines and spirits department. In that department then there were people with 25 and 30 years' service. I remember wondering how they could stand it that long. I began sorting bottles in the yard, wine bottles that had come back from various outlets. Some went back to the manufacturers but others were put into these big wooden boxes called coffins. That's when I began to take an interest in wine. There were all these fancy labels and they intrigued me. I would go into the bottling department and ask 'What's this, then?' yet despite their length of service a lot of them had no idea. They couldn't even say the names on the labels properly – they'd come out with 'Neen-steiner' or 'Leebfrowmick', the latter always followed by the question 'Is that *Irish* wine?' I got really interested and I wanted to be able to pronounce these wine names correctly. I got talking to Harry Ridley one day and he said 'Do you want to go on a wine course?' Naturally, I said yes. So I was the first on the shop floor to go on a wine course. It took place in Nottingham. In those days German wines were the biggest sellers - we sold more than fifty different brands. So I was really the first person to take a real interest in the wines we were bottling, and soon Sam Sills and I became fascinated with the trade. For working class lads, being able to taste, appreciate and recognise wines which were being drunk by the higher elements of society was quite something. Of course, in those days we were all known simply by our surnames. It was 'Bartle this' and 'Sills that'. But we rarely saw or had anything to do with the staff from the offices, other than when a director or a manager appeared among us. The people we did see and lived in dread fear of were the Brewers.

They could – and would – sack you on the spot at the drop of a hat. They had this power. I remember once when Edward Deakin, who was Head Brewer at one time, appeared in one corner of the brewery yard. I was sorting bottles at the time. He wore a white coat and a straw boater hat. My mate Dave was smoking. Dave spotted him and cupped his smoking cigarette in his hand behind his back. Deakin came up. 'What's going on here, then?' We just indicated we were sorting bottles. Deakin went round Dave's back and got hold of his cupped hand and squeezed it shut so that the cigarette was burning him. Suddenly Dave yelled out 'F*****g 'ell!'. And Deakin snapped 'Put that fag out! If I catch you again you're *out* – right?'"

"I remember when we cleared the old bottling room out and the builders were in to convert it into our new computer room. We always used to use the same local builders – Rowses. These blokes were in there having a break and they were already a week behind. I remember coming down these iron steps from that room – they're still there today – and I saw below me this straw hat coming up. I ducked behind a door to hide and Deakin stormed past. I could hear the exchange between him and the builders.
'You men – what the f******g hell's going on in here!?'
One of them replied 'Nothing, Mr. Deakin, we're just having our break.'
Deakin snapped back 'F******g *break?* You're doing *nothing!* F*****g *break*?!' The foreman didn't get chance to respond. Deakin just said 'F*** off! Go on! You can all F*** off *now*!'
And so they all packed up and went home and the next day a new gang of builders arrived. Yes, the Brewers had power alright".

By about 1967, John had acquired the skill Sam Sills would learn a few years later – that of breaking down bulk spirits such as rum and gin to 77% proof by adding ionised water. He was taught this craft by his then boss, Dennis Bradley. It would stand him in good stead. Over in Hull, the future unsuspecting subsidiary of MBC, Hull Brewery, had begun delivering beer by tanker. The tanker vs. barrels move was a pragmatic one based on commercial common sense. Mansfield Brewery added extra sophistication when they began to assemble their fleet of vehicles – with tanks which were glass-lined. In those days, the busy working men's and miner's welfare clubs could easily sell over five barrels of bitter in a weekend, which meant the cellar men or bar staff were constantly having to change barrels. By having a large tank in the cellar, this inconvenience was eliminated. It also meant an easier job for the dray men who, rather than have to man-handle heavy barrels, simply connected the beer pipe from the tanker to the inlet and pumped in however much beer the establishment needed. The only difference between the barrels and the tanked beer was that the tanker ale was not pasteurised because it would be sold quickly. For this and other reasons, some drinkers claimed it to be a better drink, its taste being much closer to that of traditional ale. Because of the new – and less physically demanding – task of the drayman, many lads in John's bottling department applied for tanker duties, which eventually resulted in him being one of the few employees left who could tackle the job of 'breaking down' spirits. His genuine and growing interest in wine also endeared him to Harry Ridley. John was eventually made foreman in wines and spirits, and soon had his own team of young lads who would go through the same initiation he had experienced. Eventually the other great wine aficionado in the wines and spirits office, Les Pearson, retired, to be replaced by Dave Long, who soon took up his own interest in wine. John's interest in wine was equally matched later by Sam Sills when he joined the department.

"When I was 18 I was moved out of wines and spirits because they didn't want to pay me a man's wage, but I wasn't sacked. So I was sent to work in bulk beer processing for nine months with a man called Fred Hensleigh. Being involved in the beer department you were more considered to be a bit of a 'nobody'. So when one of the other lads left the wines and spirits department I was asked 'If we give you a bit of a pay increase would you go back into the wines and spirits department?' To be honest I'd have gone back in there for a *drop* in wages. It wasn't just the drinking involved – I was genuinely interested in the job and in wines."

John Bartle has his own stock of Harry The Wine stories.

1991: In Vino Veritas: left to right: Julie Else, Harry Ridley, Betty Ridley, John Else and John Bartle at a Wine Club Evening in the Cellars.

"Harry knew his wines. I remember one time we were in the wines and spirits department about to bottle 27 hogsheads of Burgundy, and Harry arrived back from one of his buying trips to France. He wandered in and we asked him if he'd had a good time. 'Oh, yes, super….' And whilst he was talking to us this lorry turned up outside with 25 hogsheads of wine. A hogshead is 54 gallons, so that's a lot of bottles. Well, Harry hadn't told us this was coming and when we asked him about it he said he couldn't even remember ordering it. He was probably drunk at the time. Well, we simply had nowhere to put it and Harry, realising that there was no kind of advance warning or paperwork for this delivery, said 'Christ! I'll get the sack for this if Tony Scruby finds out..' Of course, Scruby did find out and Harry didn't get the sack. Scruby came to me and said 'Get all this lot bottled, John.' So, using ordinary corks and no labels we eventually bottled it all and it was stacked up to the rafters. Eventually we found somewhere to store it at the old mineral water department off Dame Flogan Street in Mansfield. They'd moved that department to Great Central Road so we were able to clear the cellars out and we put Harry's unscheduled shipment down there – and forgot about it. In stock terms, it didn't exist – it was 'lost'. Eight years later when we'd run out of the Nuit St. George they told us to fetch some of that 'lost' wine down and it turned out to be one of the finest vintages of the century – at the White Post they couldn't get enough of it. We were sticking all sorts of labels on it but it was such a quality wine. Harry, as ever, had known what he was doing. Compared to buying a bottle of wine today – say, Chardonnay or something like that, where you're drinking probably last year's vintage or something rarely more than 2 or 3 years old, in those days a wine had to be bottled for anything from five to eight years before we sold it. Wine buying then was a more sophisticated operation than it is now. Harry did something similar in the 1960s with vintage port. He'd gone out to Portugal and bought two pipes of vintage port – a pipe contains 96 gallons – they're called 'pipes' because they're long, tall barrels. We had a job even getting them into the building and again we hadn't known they were coming. We had to use the lift and slide the pipes in on their side to get them up to the top floor where we would bottle them. We had no proper equipment up there to bottle this stuff – every bottle was filled by hand and corked on a single corking machine, all by hand. We had to label and stamp it all by hand, and once it was all bottled we laid it down in the brewery cellars, where it remained for years.

Ten years later we opened this port and it turned out to be the best vintage ever. Mansfield Brewery's Chairman then was Robin Chadburn, who had three children. He ordered £1,000 worth of wine for each of his children, and £1,000 worth for himself. He'd told Harry that he wanted to buy the wines as an investment, so Harry duly made sure that he'd get the finest wines he could, including Champagnes, various wines and, of course, that port. Mr. Chadburn had the bulk of that '63 port vintage. 20 years later he sold the port at Sotheby's for £30,000. If there's still any of it around now it would bring £1,000 per bottle. Yes, Harry Ridley knew his wines."

Sam Sills recalls that there was a more deeply felt mutual respect in those days between the higher management and the workers.

"When I first started at Mansfield Brewery the last thing on my mind was that I would ever work anywhere else. Yes, the wages weren't brilliant – but we had a drink every day and at least we went home happy. We used to say it was a bit like living in Russia. But when we had our first change of management – when Dennis Foster came in – it was a shock to us. Robin Chadburn, for instance, always took a genuine interest in our families. Claude Chadburn was the same. My father had cancer and he had six months to live. There were Sunday visits in those days to the Brewery cellars, when people could come

and take a look around and see how we made the beer, and of course, sample the brew. Michael Parsons, who was Head Brewer at the time, said 'Your dad enjoyed his Sunday visit, didn't he?'
I replied that yes, dad had loved every minute of it, so Parsons said 'Well – you just bring him on every Sunday visit you like.' That was the kind of people we were working with back then. When my dad died, the first wreath to arrive was in the shape of a large 'M' – from Mansfield Brewery. That's how they were – they took care..."
Sam Sills also recalls the wonderful bonus of being a brewery employee – the 'daily allowance'.

"When I was a young lad one of my first jobs every morning was to fetch the daily allowance for the office staff from the loading stage. I used to fetch two and a half crates of half-pint bottles – 56 bottles. I put two bottles on everyone's desk in the office – that was two bottles for every member of the office staff from the Managing Director down. If that had gone on, by the 1990s we'd have probably needed two trucks and several forklift drivers to keep the growing office staff supplied. I remember one bloke who worked with us had the nick-name of 'Clinky'. He would buy a pair of wellingtons for working in the yard which would always be two sizes too big, so that he could slip two bottles of barley wine down each wellie. When he walked across the yard you could always hear him clinking – hence the nickname.
The reps – those out on the road, got a weekly allowance. Like most people they'd take it home, but some of us drank it on the job. The origins of the allowance went back to the days when the men who worked in the heat of the coppers used to get so hot that they stripped down to their underpants. They were given four bottles of nut brown ale per day. Eventually the men on the drays got two bottles per day. Of course, when the office staff started getting their allowance they got 'posh' beer – pale ale or strong – we got nut brown. The office staff used to get a turkey for Christmas – we used to get a chicken. We used to start at seven in the morning and some blokes drank sherry to start the day – a big glass with two raw eggs in it topped up with sherry. There were some dray men, such as Henry Goodwin, who

drove a van, after starting deliveries at 7 in the morning, at their first call would be offered a packet of fags, a drink or a breakfast. Henry used to have a breakfast of two pints of nut brown. They'd make maybe eight deliveries in a day – two pints a time – then come back to the brewery and drink their allowance. Then Henry would go up to the Victoria pub to finish the day off. It's estimated that Henry downed up to 30 pints in a working day, yet you'd never know he was drunk. The only giveaway was when his trilby would slip onto the back of his head. He never slurred or gave any indication he'd drunk so much.
I went out with him in his van once and he said 'Now then me lad – as you know, I like a drink – but don't *ever* try and keep up with me…'
I remember when the lorries used to arrive back in the yard. One night a driver backed his wagon so skilfully into a space where there was hardly a couple of inches left on each side between the other trucks. When I opened the cab door the driver was so pissed he fell out on me. If ever you went out with Henry when he was sober, he was a terrible driver then. He'd sit at the traffic lights until there wasn't another car on the horizon, then set off, probably stall the van half way across a junction. But when he'd had a drink everything went as smooth as clockwork. I know it seems remarkable looking back, but apart from the scrapes Harry Ridley had we never experienced an accident, except when one driver swerved into a ditch to avoid an on-coming vehicle".

Of course, drinking and driving was soon carefully analysed and the national statistics made grim, long-overdue sense and that era was bound to come to a predictable end. Such behaviour at work today would be deemed bordering on criminal. Yet this was an accepted part of brewery culture, and no doubt had its deep roots set firmly in the old horse-drawn dray days. Back then those well-loved and intelligent beasts instinctively knew when the day was over and could make their own way back to base, much to the relief of their sozzled drivers. As we've seen, the social side to brewery life was an integral part of the job. John recalls another aspect of this;

"Every year we went on excursions. In the wines and spirits department we would write our suggestions as to where we might like to go and put them in a box. In those days the Great Central Railway ran through the middle of the brewery on Great Central Road. So we would hire a whole train for our excursion. The guard's van would be filled with beer. You got two pints going out and two pints coming back. Our department of course had their extra supplies, stuffed under the seats and all over. I recall a trip to Blackpool one year and we were all bladdered before we got there. Of course, then we'd spend all day in Blackpool having a few more. Coming back some of the young lads would leave it to the last minute and we'd all be leaning out of the train windows cheering them on as they ran as fast as they could to get on board for the trip back to Mansfield. Some breweries were known as 'dry' breweries – although they had excursions, they didn't go in for allowances. Mansfield was a 'wet' brewery, although of course you still weren't supposed to drink to the extent that we all did. I remember we also got a jug of beer every Friday – about 8 pints. That jug used to come around on Friday afternoon and we'd drink it just before we went home. Another character at the brewery was one of the accountants. He'd come in and drink up any of the beer we'd all left. He'd been a good boxer and if he got mad you stayed out of his way. I can still see him now, shadow-boxing in his office. He'd lived all his life with his mother, who kept him. The odd thing was that he never opened one of his wage packets. He kept them all in a big tray, unopened. He always had gold pens, a gold watch, and he'd keep a roll of money in his back pocket. We used to ask him 'How much have you got today then?' And he'd whip this roll of notes out and say 'Enough to choke a pig!' Then he'd put it back in his pocket and walk off. When his mother died he became more eccentric. His auntie took him in. When his auntie died he really did go to pieces. He began scavenging in bins – someone would throw an orange or something away and he'd go and retrieve it and eat it. Eventually this behaviour caused the office girls to start complaining. Poor bloke got demoted and put in charge of the stores. But then the men started complaining because he'd be scavenging on the rubbish tip every morning for sandwiches that they'd thrown away. This was all to do with the fact that he'd not spend his money. When he started going around drinking all the dregs left over and stuff left over in the pumps in the cellar, then it was time to let him go. He was retired early."

Not everyone – especially those of a more tender age or disposition – could handle the drink available. Sam Sills remembers one occasion when two lads – who shall remain nameless – fell acutely foul of a fine batch of sherry.

"This lad had called us from the sherry store and he was in a bad way. When we got down there we saw that there was brown stuff dribbling from the corners of his mouth. We tried everything to get him sober. The Head Brewer, Michael Parsons was called. He said 'How much have you had?' The lad said 'Three glasses'. Parsons said '*Three glasses?!* That's not drinking – it's *stealing!'* He was in a bad way. They took him up to the ambulance room but eventually he was taken to hospital to have his stomach pumped. The same thing had happened with one of the other lads. Anyway, the next morning, these two turned up for work at half past seven looking pale and sickly after their experience, and they were called up to Parsons' office. I think they thought they were about to get the sack. But Parsons had poured two large glasses of sherry out and they were standing there on his desk. He said 'Here – drink this'. They said 'We can't – we feel sick just looking at it.' Parsons said 'There are only two people ever opened a bottle in that sherry store and that was you two – you'll not do it again, will you!?' Mind you, I'll never forget what Mike Parsons said – 'We've all been there, haven't we….?'

John Bartle affectionately remembers another lover of the grape and grain.

"There were those who didn't drink and there was a definite split between them that did and those who didn't. I remember another employee, Alf Ansell, who would never drink in his own time – only in the brewery's. I recall Alf well because when we'd finished work and would be crossing the road to go to Fred's Kellar he'd always pass us

in his flat cap, puffing his pipe. He'd say the same thing day in and day out, 365 days a year – 'She 'ain't been a bad'un again, has she?'
Albert Gregory probably drank more than any of us. He'd rattle the shutters at seven in the morning and come in to have his 2-yolk breakfast – a large sherry with two raw eggs. He had his own little painters' shed up on the roof. At eight o'clock he'd come down to us to see what we had available that day and have a drink with us. Then he'd do the rounds, a pint here, a glass of wine there. Between all this he'd be supervising his painters. He'd bring a jar of mussels in or some roll-mop herrings. He was a wonderful character."
By the time this trip down memory lane drew to a close Sam Sills had served us some delicious seafood from his pub kitchen, and as if paying homage to those days gone by, unlike then, today always with a careful eye on the legal limits, we'd drunk another glass of his good beer. On the way back to Mansfield the thought arose that, surely, there couldn't be more stories like John and Sam's. But there are. Brace yourselves….

1950; Jim Wain adopts the classic Drayman's 'ready for action' pose whilst warming his shoulder on the radiator of his new lorry.

Spring 1982: Left to right: Brian Oldham, Sam Sills, Gordon Jervis and Harvey Johnson displaying his passport and ticket for his first holiday in 40 years with Alf Ansell (front) and Fred Hensleigh.

Chapter Three
Brushes, Barrels and Kegs.

*"I was a time served painter when I joined in 1950.
Yet painting became the least of my jobs in some ways;
after I'd done some lagging (insulation) around
the new beer tank, the Chairman, Robin Chadburn,
asked who'd done it. My name came up.
'Right,' said Mr. Chadburn, 'in future Albert Gregory
will do all the insulation…' At times, working at the
brewery was like building the Ark…"*

Albert Gregory, Painting Supervisor 1950-1983

The Last of 'The Mafia'

Today we're in the front room of a smart bungalow in Sutton-in-Ashfield to meet some very special Mansfield Brewery workers. They've turned out in their best suits to tell their stories. This is not because that exotic animal, a writer, has come to call. This trio of smiling raconteurs are all retired, long service men who come from that generation before 'smart casual' was the dress order. They are men of respect who recognise a sense of occasion; putting their memories into print is an important event. On the table there's a handsome buffet of Stilton, pork pie, crackers and cake, with the additional bonus of a bottle of Famous Grouse and a bottle of Gin. So with all the ingredients of an epic assembled, we begin. First, some introductions.
Albert Gregory, at the time of writing, was a spritely 87 year old who joined MBC in 1950. Sadly, he passed away before seeing his words in print.

His initial duty was that of Painting Supervisor, but as we shall see, his job covered much more than brushes and paint. Brian Oldham began his days at the brewery in 1960 as a cooper. When the ancient art of barrel making receded with the introduction of metal casks, he also had a variety of jobs, retiring in 1994 as Brewhouse Supervisor. Gordon Kendall remembers the days before metal casked beers. Gordon served 50 years with the company from 1946 until 1996; he too had a variety of jobs during that time.

"When I started in 1946 the big seller then was mild. It was barrels then. As drinkers, you started off on mild, then graduated to a pint of mixed, and eventually in the pub you might get around to a pint of bitter – if you could afford it. My wages then were 26 shillings per week. My first job when I started was bottle sorting – I had to sort all the bottles out which had come back from the pubs and put them into the right cases. They used to move you around – on to bottle washing, then downstairs onto bottle filling. It was a stricter regime when I joined Mansfield. There was a dividing line between the yard and the office and we didn't cross it. They called me – or at least, the Head Brewer 'Tricky Dicky', from back then did,– 'the last of the Mafia'. Tricky Dicky insisted that everybody who was there from before his time were 'The Mafia'. When I was there in the 40s they'd only just stopped using horses for deliveries. I once remember one of the labourers telling the joint Managing Director, Edward Deakin '*when you sacked all the horses, you kept the asses'*….that bloke didn't stop long…

I had a break from the Brewery for a couple of years for my National Service. They sent me to Egypt – at least it was a bit warmer there than Mansfield. But my job was still there at the brewery when I got back. We worked a long day – 7 in the morning until five, and Saturday mornings 7 to 12. Lads couldn't work overtime then – you had to get special permission. There wasn't any kind of initiation ceremony for me, but there was for the coopers who made the barrels. Those lads were covered in oil and grease, stuck into a barrel and rolled around the yard. When casks came in and everything became metal, the coopering trade vanished. There's not many coopers left in the industry."

Brian Oldham knew all about coopering. This was his trade.

"I come from four generations of coopers. Three brothers were in the trade, My father was a cooper. It seemed automatic that I'd follow on. Like most trades you served a five year apprenticeship. When you reached 21 and came out of your time you went through the initiation ceremony. You'd have made your 54 gallon barrel – called a hogshead – and the other coopers would put you inside the barrel and cover you with all sorts – old beer, shavings, a bit of soot, whatever. Then you'd be rolled up and down the yard in the barrel. That was the initiation ceremony. I remember mine well.

Of course we weren't making brand new barrels all the time – because 90% of our day was taken up in repairing old barrels. Some of these were anything from 40 to 50 years old. Often so many of the old barrel's staves were replaced over the years that none of the original wood would be in there – but the barrel always retained its original number. In those days so many things were made of wood so the cooper's job covered more than barrels - coopers would make buckets, butter churns, milk churns, all sorts. Our foreman cooper was Ernest Chamberlain in those days. I always remember our Christmas parties when Mrs. Chamberlain would cook a leg of pork for us. Everyone turned out for a taste – the Chairman, Mr. Chadburn, would come down to the cooper's shop for his bit of pork.

With the introduction of metal casks our trade faded out, but the company always found us something else to do…"

If anyone can be said to have been from 'the old school' at the brewery, then Albert Gregory is the man.

"I was already a time-served painter when I joined the brewery on May 1st 1950. I was 30 years old. My job at the company was Department Supervisor. I ruled with a rod of iron. I took the company's painting scheme away from old, dull brown lime wash walls to a new system of painting. I was a specialist in two kinds of paint. This was nothing to do with any regulations – you could say that in the painting occupation I made the regulations. We had our own paint shop and I had three time-served painters and there were the estates painters who looked after properties out in the brewery's territory. When the weather was bad in the winter, and they couldn't work on the various

in the late 1960s I visited his office and he offered me a cup of tea. I refused it, saying 'Well, I had one once and it turned my insides rusty...' He said 'You're the first man to come in here and refuse a cup of tea....' But it was all a question of being straight with each other. Of course, they had another side to them. Deakin liked to come down to my paint shop and talk. The paint shop was full of very valuable stock and it was my responsibility to keep it locked and secure. Even if you were only going across the yard to the toilets, you had to make sure it was locked – it was a tempting place to pilfer, because the materials we stocked were of the highest quality.

Deakin would come to talk to me about life in general. He'd stand in the paint shop doorway with the door open, one foot on the step, and say things like 'You and I, Albert...my, we've seen some changes, haven't we....'

And so the conversation would go on. On this particular day he seemed to stand there a long time and we talked and talked until he eventually left. When it was knocking off time, I got ready to leave, went to lock the paint shop door and was horrified to see that the key was missing. Now this was serious – all the locks in the brewery were Chubb locks and if one key went missing they all had to be changed for security purposes. I was in trouble if I didn't find that key. So I had no choice but to turn everything out of the paint store and look for it. You can imagine what a sweat I was in. I turned everything out of the shop looking for these keys. We had a large drum with a strainer on the top into which we'd empty all the paint dregs. It left all the paint in a thick residue which we used to line rain guttering – it had turned to what we called 'fatty' and it was very rainproof. So I even delved about in this thick paint sludge looking for that key. No result. Then I retraced every step I'd taken that day throughout the brewery. No good. It was now almost six o'clock and I thought well, now I'll have to go and face them in the brewing office. So I appeared in the office and announced 'I'm ever so sorry, but I can't find the paint shop key....'

They were both there – Philip Baker, the Brewing Director, and Edward Deakin. They could really put it over when they felt like it. They looked serious. Mr. Baker went through his pockets producing keys saying 'Is this it – is that it?' and then Deakin opened his desk drawer and if you'd seen the number of keys he had they would have half filled a bucket. They could both see by now that I was at the end of my tether. Then Baker produced a key and said 'Here's your key, Albert.' Although I was relieved I realised that they'd done me – it was their joke. But to me it hurt, because this was too serious. At that time it was the period where we were about to get our first Mansfield Brewery beer tanker. Two weeks after the missing key episode, Mr. Deakin came up one morning bearing a small replica – a model – of the new tanker lorry which would shortly be delivered and go into service. He said 'I'd like you to paint this model in the brewery's colours, and I want it finished by next Monday morning so that I can put it on Mr. Robin Chadburn's desk as a nice surprise for him to show him what the new tanker will look like'. With me at the time was another painter, David Rhodes, and I turned to him and told him 'Right – I want this model painted in the brewery's livery.' The next day it was done and the day after that Deakin came up and asked how it was going and I said 'It's finished.' He asked where it was. I took him over to the bench and he looked shocked. 'That's not it!'

I said 'Yes it is – you asked me to paint the model for Mr. Chadburn and here it is.' He exploded, shouting '*That's not it!!'*

Of course he was right – I'd sent a lad out to the shops to buy a toy vehicle – any toy vehicle – and in this case I think it was a jeep – and had it painted in Mansfield colours. The original model was done and finished but I'd hidden it in a place in the brewery where no-one could ever find it. I knew that that night Philip Baker would come down to the paint shop when I'd gone and search for the real model, so I got some white cotton and attached it from a piece of timber to the door handle so that if anyone had entered the paint store the cotton would snap and I'd know that they'd been in. Sure enough, it happened. The next morning when I went into the shop the cotton was snapped but on my bench was a slip of paper which read '*Albert – black cotton is better than white…'* Later on that day Mr. Deakin asked me to go to his car as he wanted me to go to his house with him. This wasn't unusual, as we used to do painting work at the directors' houses. We arrived at his place and Mrs. Deakin said 'What's Albert doing here?' and he replied 'Well… I thought you'd like to see him.' He then spent some time showing me his chickens – he kept battery hens. He then showed me his guns – he did a bit of shooting – and I thought this was all a bit curious until later I realised that this was a ruse to get me away from the brewery whilst Mr. Baker gave the paint shop a more thorough search to find the missing tanker model. They never found it and on the Monday morning just before Mr. Robin Chadburn was due to arrive, I appeared with the real model. Deakin and Baker just looked relieved, and said quietly 'Thank you, Albert…'.

Roll out the tanks

❖ ❖ ❖ ❖

Draymen, 1930: Left to right: A. Lucas, F. Groves, H. Boddy T. Goldsmith

Chapter Four:
To Hull and Back

1972: Hull Brewery's finest: a young John Goulden

The Memories of John Goulden, Area Manager for Hull Brewery, North Country Breweries and then Mansfield Brewery.

"I was sixteen when I started with Hull Brewery. I can't remember what my title was but it was really impressive. It was something to do with statistics – so let's call it 'buggering about with figures'.

When I first left school I was interested in going into accountancy without really realising what accountancy was about. But other than games, figures and maths at school had always been my favourite subjects. So I thought this could be interesting as a job.

I was one of the last wave to do National Service – those of us born in 1939 were the final ones to be called up before conscription was ended. I went into the RAF and was sent to Cyprus. I was one of the lucky ones because I really enjoyed myself. It was brilliant. The Commanding Officer of our small Unit was a Squadron Leader who was an absolute football nut. He excused about fourteen of us for games – football – and for just over two months of our National Service we never even went into uniform. So I was in Cyprus, in the RAF, but not in uniform – and we were living a life almost as professional footballers.

After this pleasant interlude I was back at the Brewery. I was moved from statistics into stock taking in 1963. Then I became a kind of figures Jack of all trades. They asked me to take a general position where I was going to be the holiday relief figures person, a Tom, Dick and Harry role in the wages department, etc., anywhere involved with figures. I did that for about two years and I enjoyed it, then went back into stock taking as head stock taker until 1972 when Northern Foods bought Hull Brewery out and we became North Country Breweries. Prior to 1972, the old Hull Brewery, as was the case with many smaller regional breweries, operated with no actual recognisable structure of management. The directors then were also the managers. The directors were the people in power, men such as Colonel Cooper, who was the Chairman, Major Gleadow, Colonel Black, Leslie Cooper and Geoffrey Trueman. The only other people with power were the Transport Manager, Jack Pickering, who had a considerable amount of power and the Company Secretary, Henry Jephcott. But the man who had the most power seemed to be the Head Brewer. Oddly enough I can't remember his first name but his nickname was 'Schnozz' Hughes. He was a damned good brewer but a nasty little man who wielded immense power to the point where he could – and did –dismiss people. It was a case of 'Put your coat on – on your way home – that's it – you're finished.' However, when Northern Foods took over and we became North Country Breweries they formulated this grouping of district and area managers and at last some of us had some power inasmuch as we were given areas to be responsible for. This meant the hiring and firing of staff, appointing licensees, and the profitability of your area. Did I make any cock-ups at this time? Frankly, yes – some real humdingers, especially in the appointment of licensees. And there were some interesting stories during the stock taking days. For instance, one of my colleagues in those days, a lad called Ken Charles, who has unfortunately passed on now, hadn't been in the stock taking department very long at all when he was sent to do a stock take at a pub called The Dairycoates on Hessle Road. I think it was his first solo stock-take. Anyway, he motors off up Hessle Road in his little car, parks outside the pub, gets his stock

Yorkshire & England cricketer John Hampshire presenting John Goulden with prizes at a Hull Brewery function.

board out and in he goes. When you start a pub stock take like this you always begin with the spirits on the shelves, the optics, etc. So he commences by asking the licensee what did he think he was doing stocking all these products which were brands handled by Moors and Robsons, which was another, smaller independent brewery in Hull. It only dawned on him after a few minutes that he had gone into the wrong pub. He'd gone into the Locomotive, a pub about three doors away from the Dairycoates, and there he was giving the poor landlord all this grief when he shouldn't even have been in there in the first place. He had to walk out with egg on his face. Another good friend of mine, Rod Prosser, had his first stock take also on Hessle Road at the Star & Garter. He completed his stock take as far as he was concerned, closed his book and left. Anyway, he was waiting at the bus stop on Hessle Road when the licensee from the Star & Garter came running after him, waving his arms. Rod had only forgotten to do the pub's cellar – and beer sales in those days accounted for 80% of a pub's turnover. In a Hull Brewery Pub's cellars there were enormous porcelain jars which varied in size; the largest was 108 gallons – a three barrel size, there were 90 gallon jars, 72 gallon jars, 54 gallons, 36 gallons and 18 gallon jars, depending on the size of your pub. In those days Hull Brewery delivered their beer in tankers. It was only the little country pubs who had the cask beer then. Another mate of mine was completing a stock check in the Monument, a lovely little pub which used to be down Whitefriargate in Hull; the kind of place where you could sit and have a good chat – a 'talking' pub. Anyway, he's doing the job and there's a pitter-patter of high heels comes down the stairs and the licensee's wife ran off into the street with the week's takings – she was doing a runner. So it was pointless checking the cash that day – there wasn't any to check. But they caught her eventually. But in those days – I don't know if this was the case with Mansfield Brewery – we didn't go through all the legal rigmarole of prosecution and such – the misbehaving licensee was simply dismissed. In this vein, another of our pubs was aptly named the Crooked Billet, Ings Road. It was a good name because four licensees there were all crooks. One of them actually deliberately banged himself on the head to pinch a week's takings – making it look as if he'd been robbed. In those days they only banked their money once a week on a Monday. We had two ex-policemen employed as cellar inspectors. I remember them giving this guy the shake-down in an attempt to get the week's takings back. Taking stock sometimes meant that you could cover three smaller pubs in one day, for instance the Monument, the Bonny Boat, which was a bargee and fishermen's haunt, and the Kingston. Some of these landlords were crafty and had a clever scheme going to mask the stock situation. For instance, you might find at the Bonny Boat that the stock all seemed correct but two or three landlords were working a flanker. They'd make sure that when we checked the pubs in sequence that extra stock would be moved from the first pub to the second so that everything tallied. Whilst we'd be upstairs to do the books, after checking the bar, they'd be busy downstairs moving the stock on to the next pub we'd be visiting. We soon cottoned on to this and two stock takers would appear simultaneously, one at the Monument and another at the Bonny Boat – so there'd be no 'portable' stock moving from pub to pub to keep the balance right. At the Providence in Spring Street we had a lovely licensee called Kate East who worked into her 70s. Everyone loved Kate. Kate had a cleaner who used to go in every day on her hands and knees and scrub the floor boards. They were beautifully cleaned, bleached white. This cleaner had the most enormous callouses, but she also worked into her 70s because, like Kate, she loved her job. I fondly remember when we used to take stock in that little pub during winter time. These pubs didn't have the kind of heating they have now; the Providence was only warmed by its coal fires. But at 9.30 in the morning a little baker's van used to come and stop outside and we'd sit in that pub with Kate eating these delicious warm bread cakes, lathered in butter – absolute nectar."

"Some of the fiddles that used to go on in those days used to make licensees quite a nice side-income. In later years, one of the main ones became known as 'passing off'. This involved the pouring of what we regarded as inferior products into branded bottles and passing them off as what was on the label. Indeed, lots of pubs then used to sell an awful lot of sherry, and at Hull Brewery we bottled our own. Our main brand was actually a Bass product – Bertola Sherry. We used to bottle it and a lot of our pubs stocked it – but they were also filling the bottles with QC – a far cheaper brand, and making a tidy little living from these activities. They simply just kept re-filling the same empty bottle. But despite this we didn't get too involved or upset. We took it easy – as long as the Brewery got what it wanted from its pubs, what little sidelines the licensees were involved with were in general overlooked. How times have changed...."

1978: Colonel Trueman, Wines & Spirits Director presents a cheque to Don Thompson Darts League Secretary as John Goulden and Terry Crockford (both far right) look on..

"I remember a landlord at the old Victoria pub on Hull's Hedon Road, down by the timber docks. A lot of Finnish vessels used to come into the dock there and the Finnish sailors drank a particular brand of bottled lager which we sold then – I can't quite recall the brand name. But it sold at around one shilling and sixpence – amazingly just over 7p a bottle by today's currency – yet when the Finns came into the pub the landlord was selling them this lager at two shillings and sixpence a bottle. This was a hefty mark-up which earned him a tidy little fortune back then. At that time the pub was a managed house. However, I recall that when he came to us to take his tenancy and pay for his in-going, this included a sum for his fixtures and fittings. He paid in cash with a mixture of five pound notes, pound notes and the old ten shilling notes – and they all absolutely stank of mothballs. A lot of these landlords who were making these side-incomes used to store their money not in the bank but in old World War II ammunition boxes. You could buy these at the Army & Navy Stores. They were great to store cash in with a handful of moth balls – after all, you didn't want the moths eating your ill-gotten gains!"

"Then there was the couple who ran the Star & Garter on Hessle Road, Peter and Clarice Bevan. Peter wasn't the licensee – Clarice was. So when one of the pubs most loved customers passed away, Clarice was too busy as usual to attend the funeral, so Peter was delegated to go. He attended the church, then later in the cemetery stood by the graveside sobbing his heart out. None of the relatives recognised him, and asked him if he was a relative. After he'd explained that he was from the Star & Garter on Hessle Road, they were even more puzzled. It was revealed that Peter was, in fact, at the wrong funeral in the wrong cemetery; the deceased customer was being buried in Western cemetery and he'd turned up at the Eastern cemetery. ..."

Now that we're living in the 21st century there are stories we feel confident to include here which may have been a trifle too risqué a couple of decades ago. Listening with rapt attention to John Goulden's vast experience working for Hull Brewery, North Country Brewery and Mansfield Brewery was an encounter filled with that familiar feeling we had become so used to in talking to ex-MBC workers; the sheer affection for not only the job itself, but for the whole community of pub and brewery employees. No doubt many of the participants in these original recollections are now in that great bar-room in the sky, but if they're not, or if you are a relative or friend from those halcyon days who recognises a character here, then their story via John is told not as an intended exposé but as an extension of that genuine fondness for all the great larger than life employees who served the Brewery so well. They did so in a less avaricious age long before rainbow-coloured alco-pops, when beer was a savoured pleasure, not drunk straight from the chilled necks of designer bottles to fuel a late-night binge. Over our delicious filled potato skins at the Green Dragon at Welton, Brough in East Yorkshire, John Goulden's memoirs continued...

"One of the best loved characters on the Hull pub scene was Ron Stothard. Possibly apart from Johnny Whiteley, who remains a revered Hull figure because of his connection with Rugby League – and he was a lovely man as well – then Ron Stothard was one of the best loved landlords of his day. He started his career at the Bean, went to the Commercial, then the Elephant & Castle, Blue Heaven, Red Hell, and then the Polar Bear, his last pub. Blue Heaven and Red Hell were the Anchor and the Crown respectively – docker's nicknames for pubs stuck in those days so that, for instance, few referred to the Anchor as such – it was Blue Heaven for years when we had the docks. Ron did the occasional spell as a relief, and on one occasion he went to the Paragon. The Paragon today is known as the Sylvester, and it was known before that famously as the Hull Cheese. Back in the days it was called the Paragon it was always noted - because Hull was a very busy port in those days – that the Paragon was the haunt of prostitutes. In earlier times sailors used to say that the ladies of the night used to sit in the Paragon – and other such establishments, one leg crossed over the other, displaying their fee written in chalk on the visible sole of their shoe."

(As to the provenance of this innovative marketing device, the author's own experience in the Paragon in Hull as a young sailor as late as 1959 will have to be accepted…)

"I went in there one night after practising at cricket nets. The girls frequented a room upstairs where they went for a drink and to carry on their trade. On this particular night, two of the girls started scrapping. It was always difficult when this sort of thing kicked off, but in the days before political correctness it was no surprise for a landlord or bar man to steam in and grab one of the girls to stop the fight. This is what Ron did; he pulled one struggling girl from the other. Her legs went up into the air, her skirt flew back and….well; suffice it to say that she wasn't wearing any drawers.

When things had calmed down, Ron said to the girl

'You know, Jean, I hadn't realised that you were so 'small-made'…'.

He was a real character, Ron Stothard. Another one was a man called Rimmington who kept the Rose Tavern on Hessle Road. His wife and his mother in law also worked there. He was a nice man, a big man, over 6 foot 2 inches, and he was the spitting image of one of the great

1960s entrepreneurs, John Bloom, who made his name with the Rolls washing machine. I remember going into his pub one morning about 9.30, and there, working away, bent over in the corner of the bar, was the cleaner. I was immediately taken with how attractive she was – especially as she was wearing a very short skirt, and I was faced with this lovely derriere clad in red silk panties. This pretty, five foot two brunette had been a secretary. What she was doing working as a cleaner I don't know.

Anyway, he told me that this was his new cleaner-cum-barmaid. A couple of weeks later I got a telephone call from him asking me if he could come into the office to discuss a 'matter of some urgency'. Needless to say, I told him it was no problem. He arrived, he sat in my office and then asked me 'John... what would the brewery's attitude be to me sacking my mother-in-law?' It was an odd question, but I replied 'Well... as far as I can see there would be no problem but this is something you would have to sort out between you and your wife. There would be no comeback from the Brewery provided of course that this had no adverse effect on trade...'

He said 'Good. OK then, that's fine.' And he left. He did the dirty deed and got rid of his mother-in-law. Around three weeks later I got another phone call similar to the first. We met again.

'John... what would be the Brewery's attitude to me sacking my wife?'

I was a bit taken aback. 'Sacking your *wife?'*

He was a manager, not a tenant, so his wife was simply paid by the Brewery what we called an 'honorarium' of about £4 per week. It was a condition that when the directors appointed people to run pubs that it usually meant married couples.

'Look,' I told him, 'this is obviously a very personal problem and we have to accept that couples do split up, so if she leaves you then there's nothing you can do; once again, if it doesn't have any detrimental effect on turnover.....'

And so he sacked his wife. He didn't stay at the Rose Tavern for long. He took over as Steward with the Conservative Club in Barton-on-Humber. When I went to see him, there he was, happily installed with, you guessed it, the attractive little cleaner. I suppose it puts a new slant on separation from your wife – not 'I'm leaving' but 'you're sacked!'...."

Eventually John Goulden's managerial responsibilities were expanded to the post of District Manager. He became responsible for the profitability of tenanted outlets, free trade outlets, leaseholds and managed houses.
"My area went as far as North Cave and included the centre of Hull, the west of Hull including Cottingham and Beverley – it was quite a large area with a lot of outlets.
When the Paragon changed its name to the Hull Cheese and we were being run by North Country Breweries – around 1977 – I'd left the Hull Cheese one night for an appointment at the Star of The West. We used to work at least two or three nights per week then on visits. I met two chaps from Whitbread who were about to call at the Hull Cheese concerning the reciprocal trading agreement with Whitbread and Stella Artois. They asked me how the sales of Stella were doing I told them they were phenomenal, because the pub was moving over 90 x 10 gallon containers of Stella in a week. They found it hard to believe me, because, they said, if that was the case, then that figure would mean that this was the fourth – or possibly third – largest through-put of Stella Artois not just in Britain but in Europe. What was responsible for the eventual dip in Stella sales – and at that time Stella was a genuine European lager – imported from Europe, was the success of another great lager – Mansfield's Marksman. It was a damned good lager and immensely popular in Hull.
When Mansfield took us over I had a lot of time for them. I told them so – that I thought it was a great brewery. I loved my job, and it was the licensees, the quality of those people, who made it such a pleasure."

"An atmosphere in a pub is generated not by a brewery or a business but by the people behind the bar – and that includes the licensee and his wife. Bric-a-brac, furniture, fittings, decoration, these are all important, but it's still the people who run the pub who make it what it is. In 1977 I took over as Area Manager over four District Managers, Terry Crockford, Keith Scott, Harold Gelder and Ken Johnson. The odd thing is that during Northern Foods' period running us as North Country Breweries they didn't really 'take over' Hull Brewery as such – what they were looking for was a foothold in the leisure industry and it is a paradox that at one time Northern Foods actually made overtures to take over Mansfield Brewery. The brewhouse in Hull was past its sell-by date and the bottling plant was finished. To justify a new brewhouse and bottling facilities we needed to acquire another brewery. I think what happened was that Mansfield's chairman Robin Chadburn and Nicholas Horsley from Northern Foods had dinner together. It appears that Mansfield were on the acquisition trail, a route North Country had been on without success, and the outcome of Horsley and Chadburn's dinner was a suggestion to Mansfield 'make us an offer'. I think that Mansfield bought us out for just short of £48 million. At the time our bottling plant was shutting, we had a reciprocal trading agreement with Bass. They took our Mild and we sold Carling lager. They took an incredible amount of bottled mild. How that agreement came to an end I don't know. At that time Hull Brewery were producing their own lager called Top Score. When it was good, it was good, but the variation in consistency was unbelievable. But when we stopped brewing Top Score we went down the Continental route and began brewing Hopfenperle. Once again the inconsistency in brewing Hopfenperle was enormous. At times it was beautiful, at others it wasn't. Hopfenperle was a brand we took up from Switzerland. I remember us going over there to the Feldschlössen brewery at Rheinfelden. The brewhouse was known as 'the cathedral' if I remember rightly, and it was the most wonderful brewery I've seen – so clean – yet then again, I'd never seen a country as clean and tidy as Switzerland. A fantastic brewery with a great museum attached which showed the history of transporting the beer – horse and carts, steam driven vehicles, fascinating. Hopfenperle was a good beer, but with us it was an inconsistent one. When Mansfield took over and introduced Marksman, then we had one of the best lagers ever…"

John Goulden's career, like that of all the interviewees in this book, has been one full of interest, progress, and packed with that good-natured love of life which the brewing industry seemed to engender. There were several stories which we simply couldn't include; perhaps they offered too acute a focus on that very rare animal in the world of wines, spirits and ale – an unpopular departmental head. One comment we received from an interviewee during the early stages of recording these interviews was that the revelations of brewery workers, when lumped together without some thoughtful editing, "Make us all look like staggering alcoholics who spent most of our time drunk…". It was a fair point. In a job where one of the perks was a ration of alcohol in various forms every working day, then to a prudent reader veering towards tee-totalism it may seem that between the drinking very little work was carried out. The evidence, however, is to the contrary. Immensely profitable businesses such as Mansfield Brewery could never have achieved such market prominence without a lot of genuine hard work and dedication on behalf of the staff. The fact that what often lubricated this well-oiled profit machine was the occasional enjoyable drink is what makes the industry stand apart from others. One has to remember that brewing and selling beer and other beverages is a serious adjunct of the leisure industry. To most men and women the word 'beer' will always be associated with fun and relaxation. But even for the dedicated brewery employee, a little too much of what you fancy can sometimes produce a bizarre adventure. When John Goulden's territory expanded, he often travelled further afield to inspect premises. Some workers at his management level might not wish to tell a story like the one which follows, but this bears a close similarity to those of many such yarns told to us by others who, although realising the humour involved, requested that we switched off the tape recorder. John is made of sterner stuff, so let's round off his marvellous contribution with a tale which any bunch of blokes after imbibing one over the eight will feel familiar with.

"With a couple of colleagues one day I'd been to the Cobbler, a pub in Pontefract. Let me make it clear – we weren't driving. We were in a nice little brand-new mini being driven by a very responsible friend of mine who happened to be a policeman. We'd had a skin full.
We'd had a few pints and then ended up drinking copious amounts of Southern Comfort. On our way back on the old A 63 on the way to Hull we pulled over at an abandoned, derelict transport café. My two colleagues needed a pee. I said I didn't, but once they'd got out I thought that yes, maybe I ought to have one as well, seeing as we'd stopped. So I hopped out the car, went trotting over the gravel to where there was some long grass. The grass and weeds had been growing for many months but had been blown flat by the wind. So I went haring into it, and seconds later, I disappeared from view – ten or twelve foot below ground into an old concealed ditch. Fortunately, I stopped a foot or so from the absolute bottom, where there was a thick stream of evil-smelling goo.
So there I am, stunned, lying at an angle of about 80 degrees, looking up, having a pee. I shout for one of the others.
'Ian?!' And there's little crunchy foot steps approaching over the gravel above. This little face appears, Ian's. He shouts for Terry Crockford: 'Crockers! Crockers – come and look at this…'
More footsteps on the gravel and then a rustle and 'whoosh' and a foot away from me at the bottom of the ditch, Terry has fallen in too. He's stunned; just murmurs 'Hello…'
So there we both are at the bottom of the ditch, it was almost as if we'd planned to meet there. I started to scramble out and made my way up, but Crockers slipped down, almost up to his knees in the foul, stinky green slime. And he's got his best suit on because after all, we were out on business. He got out eventually, but he was lathered in this gunge. On the way back to Hull we stopped at Terry's house at the village of Newport. Terry had arranged to meet his wife in Hull that afternoon, but he simply had to call home to change his trousers and his shoes. Then they dropped me off and my policeman friend takes Terry into Hull but by this time his wife has given up and gone. Terry

was still that drunk that he couldn't even see the time on the railway clock outside Paragon Station. So with his wife on her way back to Newport, he goes to the taxi rank outside the station asking for a cab back to Newport. But he's in such a state that none of the cabbies want to take him. Eventually, they toss for it and the coincidence was that the cabbie who got the short straw was actually the father of the policeman who had just driven us to Hull. Anyway, they got Crockers home and his wife was disgusted with him and put him to bed. She went out that night and left him and locked him in the house. Crocker wakes up later on and, still legless, realises that he was due to go to the local pub that night to play darts. But he can't get out of the house, so opens the bedroom window and shins down the drainpipe. He walks into the King's Head in Newport and collapses. Some guys in the pub carried him home. They had to somehow get him back up the drainpipe and back through the bedroom window. Now this is really a cautionary tale on the power of drink – and looking back, had we been caught in this state we'd have been hung, drawn and quartered. But I suppose it was all in the line of duty…"

Stories of this dimension are not quite as common as one might expect, but this illustrates the folly of passing that brewery worker's boundary where pleasure ends and pointlessness begins.
"To sum up," says John, "working for Hull Brewery, North Country and Mansfield Brewery has been an absolute pleasure and a privilege. Corporate philosophy today has changed drastically, but in those days we still got the job done, and we were profitable. It was a career anyone would have enjoyed."

❖ ❖ ❖ ❖

1987: John Bartle shocks Wainfleet's innocent bystanders with a display of ferret flashing. (Trousers were provided for subsequent trips).

Chapter Five: Whippets & Ferrets

"This was the weekend when we managed, by accident, to knock down the town's statue which had stood there for over 400 years.....the coach backed into it... We all piled out, and stood it back up on its plinth. I presume it's still standing there..."
Graham Kinder, veteran of the Supervisors' Outings.

Graham Kinder spent a great part of his working life as an engineer in the Merchant Navy. As Britain's merchant fleet began to disintegrate in the 1980s, in 1983 Graham was made redundant and his trips across the seven seas came to an end.

"I saw a job advertised at Mansfield Brewery as a fitter. I had taken a job for a while with an engineering company in Chesterfield. I had an interview with Mansfield Brewery but it turned out that the job on offer wasn't quite up to the expectations I held for it. But one night not long after, I got a phone call from Mike Cleator. The job he was talking about, Senior Foreman Fitter, seemed to be the same job, but somehow the money had improved. At sea I had done quite a bit of refrigeration work, which swung it for me.

I always remember my first day at the Brewery – and a bloke called John Wragg in the fitting shop. There was the bottom fitting shop and the top fitting shop. I recall the Christmas parties, too, and a bloke called Dick Spedding – who needs mentioning in despatches because he was one of the Brewery's real characters. Dick was an Englishman who always wanted to be an Irishman. He would always sing *The Galway Shore.* He also used to sing this song over our intercoms and radios. I remember this pitch black night of this Christmas party in the fitting shops and we'd turned all the lights out except one so that Dick Spedding was kind of silhouetted in this single light, the only electric light apart from some candles we'd lit for atmosphere. Outside the town was pitch black and blowing a gale.

One of the wives, Jane Featherstone, had arrived outside and parked her car at the bottom of the car park and walked towards the building in the dark. She said it was a ghostly experience – this lone voice on the wind coming out of the darkness, singing *The Galway Shore.* She was right. It was eerie. That's something which I'll always remember. That was the kind of atmosphere we worked in.

In those days things moved quickly. The normal working hours, seven to four, often didn't count. We might do the day and then be working in the bottling plant until half past six. I remember in the early days when we were working in the bottling plant using a 3-litre bottle filler. We had a contract to fill a 3 litre bottling order for Christmas. We had a thing called a video jetter which used to put a sell-by date on the bottle, and the time it was bottled. But everything had gone wrong which could go wrong during that day. They asked us to stay behind to complete this bottling run, because once it had started it simply had to be done. I'd left and gone home, but I got a phone call from Mike Cleator who at that time was packaging manager. He asked me to come back in as the video jetter was playing up. When I got back they'd assembled a motley crew to get this job done with all kinds of people from the Brewery. At the same time, in the room next to us, the wines and spirits department was holding a promotional evening. At one point, as we were nearing the end of the bottling run at about 9.30, Sam Sills appeared from the wines and spirits department carrying this tray with two small glasses on it and a bottle of vintage port. My word, did we partake. It's nights like that which you'll always remember. Somehow we got the job finished.

I then moved on and came out of engineering and went into production as a shift supervisor. I suppose you could jokingly say that that was when the rot set in. In those days Alan Singlehurst was the Chairman of

the supervisors' social committee. We used to have these supervisor's outings organised by Alan and Mike Parsons as a thank-you to the lads for the work done. Alan was looking for a way out from organising the trips so I volunteered. It was suggested that we could have two outings per year. One would be organised by the Brewery and the other I would organise, to be paid out of the supervisors' fund.
The supervisors used to pay I think around £4 per month towards these social events. The first big one I organised with Sam Sills was an opportunity to go up to the whisky distillery in Scotland at Knockando. We all decided to meet up at the Ram in Littleworth – and that pub holds a few stories – and wait for the coach. Anyway, the bloke we knew at the Ace Coach Company, Roger, arrived at the Ram to tell us that the coach was waiting outside. It was a brand new vehicle, toilets, videos, etc. It was an overnight ride and we got through a lot of beer, arriving in Aviemore at about 7.30 in the morning. We'd had a full night of it on the coach, but everything seemed closed in Scotland. We were looking for a breakfast and in this village – I think it was Huntley - I saw a café door ajar and leapt off the coach and the lady said she could do us all a full breakfast. We needed it. Afterwards we knew that we needed to spruce up a little and decided on some exercise – a half hour's walk. We walked down the street and right in front of the Police Station was a parked car with two absolutely drunken blokes sat in it who tried to entice us with the slurred words 'Hey, see you Jimmy – come over and have a drink wi'us!' And *this* was outside the Police Station!
Unfortunately, this was the weekend when we managed, by accident, to knock over the town's statue which had stood there for over 400 years. I can't recall the nature of the statue, but the coach backed into it and knocked it over. We all piled out, picked it back up and stood it back on its plinth. I presume it's still standing there".

(Editorial note: If this was indeed the village or town of Huntley then there are two possibilities for the knocked-over monument. It may have been the Earl of Huntley, or a monument to commemorate one

of the battles between the Clans in the area. Alternatively, there's a folk song based on events in the Spey Valley entitled *Belle's Bonnie Bogie* which tells a story of a monument erected in Huntley's square made not from granite but from…bogies (the nasal substance). A stanza goes
'They rolled it roond and roond the toon…'
This nigh-on useless information seems to fit in well with the bizarre history of the supervisors' outings..)

"On that trip I couldn't get everyone in one hotel. Being twenty of us on the outing, and the part of Scotland we were in, I had to split us up into different lodgings. It was a mix of hotels and guest houses. The crème de la crème were in the Country Lodge which was outside the town. This was the day of the infamous golf experience. They were great at the Lodge. They knew we'd been travelling overnight and the staff appeared with towels so that we could freshen up, and they even provided facilities to empty the coach's toilet – it had become quite full overnight! Anyway, that morning we went around the Knockando Distillery and we started to 'sample' the product. I asked them if they had any gin – I know it sounds odd at a whisky distillery – but the man showing us around went down into the cellars and generously came up with this big, cob-webbed bottle of Bombay Gin – it was fabulous. It was after the distillery that we knocked the monument over. I don't think anyone spotted us. But we got it standing again. Roger got us to the hotel – the Lodge – and the receptionist could see all these oddballs coming in. After we got cleaned up and refreshed we decided unanimously that we would go and have a golfing competition. But this was a competition with a difference. We had on the bus a whole bag full of miniature drinks – those little one-shot bottles – mainly gin, which we'd brought along. So we decided that the coach driver would be the caddy who would carry the drinks. The place we decided to 'play golf' seemed to have unsuitably long grass. We got our golfing gear and our free, complimentary ball, and it was decided that each time we hit a ball then the coach driver would get one of the miniatures

Visit to Guinness Brewery, Park Royal, Dublin May 3rd 1974
Back row: L-R (Guinness) Albert Gregory, Sid Clifford, Harold Moore, Alf Ansell, Fred Hensleigh
Middle row: John Ruskin, Eric Jackson, (Guinness) Bob Gilliatt, Jim Wain, Brian Rose, Les Yates, Sid Tyler. Front Row: Ian Boucher, (Guinness) Michael Parsons, Philip Baker,and George Clark

Visit of
PARTY FROM MANSFIELD BREWERY
NOTTINGHAM

Luncheon

at

ST. JAMES'S GATE BREWERY
DUBLIN

Monday, 1st May 1961

ARTHUR GUINNESS SON & CO. (DUBLIN) LTD.

Menu

Smoked Eel

×

Prawns Mayonnaise

×

Roast Loin of Lamb
Mint Sauce
Spring Greens
New Potatoes

×

Rum Babas

×

Coffee

Guinness Stout
Phoenix Ale
Fruit Squash
Cockburn's Port

out and we'd drink it. By the time we got back to the hotel I don't think the receptionists had ever seen anything quite like it. The state we were in is hardly surprising when you consider that it was now 4 o'clock on the Saturday afternoon and we'd been at it since 9 on the Friday night. When the time came for our evening meal things had deteriorated further. There we were assembled around the table in the dining room and one of our party, who must remain nameless, ordered his soup. The inevitable took place; he slumped forward with his face in the soup bowl, fast asleep, and snoring so loudly that everyone wondered what this bizarre noise was. We decided we had to get him to bed. There had been a wedding at the hotel, and there was a conference room upstairs. So we took this poor bloke to his room first, undressed him and then (these are the things you do when you think you're 'having fun') we thought it might be a good idea to conceal his modesty in a sheet and sit his prone figure in the conference room where the wedding had been. We cleared off and left him and when he woke up in the morning, in this huge room, swathed in this white sheet, the poor bloke woke up and thought that he'd 'gone over' – that he'd passed through death's veil. But somehow he found his way back to his room. Another had fallen asleep in the library and woken up, opened the door to go out and nearly fell two storeys – he'd opened the fire door!"

"There were other trips, for instance Jenning's Brewery at Cockermouth, and Adnams, in Suffolk, where we managed to out-sing the Sheringham Shanty Men, much to their annoyance. We were really well entertained by the brewery's head brewer. The nearby hotel was popular with Londoners who go there for the cuisine. The bar was shut; there were about three or four women there. The manager was called Dudley, and Adnam's Managing Director simply shouted 'Dudley – the bar!' and suddenly it was opened for us. Some of the lads got talking to these girls who reckoned they were 'in the theatre' and up for the weekend. We could tell by their forced phoney accents that these were anything but 'actresses'. Anyway, that night the

Sheringham Shanty Men were doing their sing-around to raise money for the Life Boat and we pitched in and probably raised as much as they did with our singing. This was the big event in Southwold. I went around with a big box shouting 'get some money in here!'."

Albert Gregory also recalled his first Supervisors' outing in 1951, to London. This included visits to the Smithfield Meat Market and the Guildhall. Lunch was taken at a pub called The Cockney Pride. Those attending were Albert, Les Pearson, Bob Gilliatt, Ernest Chamberlain, Percy Holmes, Tom Hinchley, Les Hardy, Bill Holland, Charlie Wilkinson and George Clark. The Brewers on this trip were Edward Deakin, Philip Baker and Michael Parsons. Joining the party at lunch was Harrington Griffiths, Secretary of The Brewers' Society.

ENTER THE WHIPPETS & FERRETS

One of the innovative arms of Mansfield Brewery's market 'research' came about when Graham Kinder was involved in the foundation of an unusual pub outreach organisation called The Whippets & Ferrets Society.

"The idea originated with the idea that a group of us should go out into the trade and visit the pubs to form a kind of bond with landlords, staff and customers as representatives of the Brewery. Someone came up with the idea that this group of roving ambassadors ought to have a theme, and so we became the Whippets and Ferrets.

We got right into it. Cloth caps, waistcoats, mufflers, the full kit. But it went beyond this, we had animals made – ferrets, and a kind of 'Spit the Dog' character called 'Bozer' who was a very well-made, stuffed Whippet puppy, carried on my arm in the same way that Bob Carolgees used to carry Spit. Well, this dog became very popular in pubs. Women used to come up and stroke him as if he was real. I'd

1990: It's a dog's life: The Cock & Magpie, Old Whittington, near Chesterfield February.
Bozer the dog with a tasty ferret in the wandering hands of John Bartle

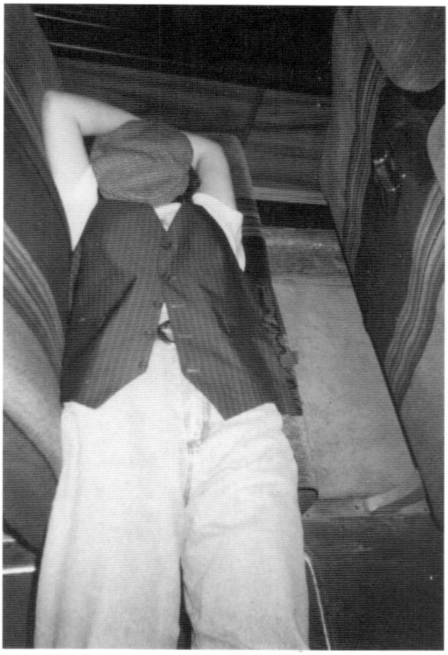

Ferret Fatigue: if you can't whippet, kip it!
Sharon Lloyd escapes from Bozer the dog on the coach on the Cock & Magpie expedition, 1990.

"You're my best mate, you are…."
John Else expresses ferret affection to Mike Cleator, Westfield Hotel, Mansfield, February 1988.

Bozer the Poseur (who reached the parts of women where other dogs dare not go), with Graham and Moira Kinder, Spread Eagle, Chesterfield October 1988.

even dip his nose in their drinks, and, to put it mildly, I got away with quite a lot with Bozer on my arm! At one point the dog was kidnapped and a ransom advert was placed in the Mansfield *Chad* newspaper. Whippets and Ferrets outings were part of the Brewery's commercial campaigns. They were hilarious expeditions."

"At one point we used to travel in a reproduction 1920s vehicle – a Criterion – and John Else was driving us one night. We left the pub and he stalled the vehicle three or four times. When we got out of the car to see what was happening, we found that someone had let all four tyres down. We had no way of pumping them back up, so someone suggested that we phone the AA, which we did. I don't know what the AA man must have thought – there was this group of people dressed in big cloth caps, waistcoats and mufflers, standing in the dark besides a 1920s vehicle with flat tyres.... but that was the Whippets & Ferrets Society experience – always something weird going on. There were fifteen members of the society. It wasn't just a management membership – it was across the board, personnel wise. We would go around the pubs simply to show the flag. We had everything – hand-painted waistcoats, ferrets, t-shirts, brass ferret key rings. We were a legend. The local people in the pubs used to really appreciate it. We went all over, up to Hull and other places. We visited every kind of pub, from the roughest to the poshest, and everybody loved us. Yes, there might have been the odd occasion when some people couldn't see the humour in what we were doing, but that was rare. The Whippets and Ferrets continued right up to the demise of the Brewery."

Another great Brewery social event was the regular Sunday visits.

For a number of years the public were invited to visit the Brewery in Littleworth on Sunday mornings to see just how their favourite brew was made. This had an extra beneficial dimension – at no cost to the visitor, you could actually try the beer. Needless to say, as Graham points out, there were those who abused this golden privilege. The public used to be shown around and discovered how we made the beer, etc., but there'd always be a hard core who couldn't wait to get to sampling a few pints. I suppose it's understandable – Sunday morning, free beer – what more could you want.

"We had some odd occurrences at the Brewery. We used to have a repertoire with the Fire Brigade and we used to go through routines with them such as gas exercises, warfare scenarios, that kind of stuff. At one time they asked me if they could do an exercise in the Brewery on gas warfare. I mentioned to the fire chief that we had a problem; we had a cowl blown off one of our huge water tanks – they were very high. We decided that we'd have to tackle getting this cowl – a kind of lid – back onto the top of the water tank when it was quiet, after working hours. So we have this brand new hydraulic device with a chair on the end which we're going to raise right above the tank. Mike Cleator volunteered to go in the chair. Well, we got him poised over the tank, just ready to lower him down, when the computer driving this machine broke down and we couldn't shift it. He was stuck up there in mid-air. It was a German machine and we're having to phone Germany whilst Mike's still stuck up there, wearing a boiler suit, to find out how to get him down. In the end, we had to winch him down manually. Mr. Parsons – as he was known then – Michael Parsons, was always keen with the instruction – 'Fred's Kellar'. If he said that, it meant you had to go to the Kellar for a drink. You'd open the door and a voice would come out 'Hey up, Graham, *where've you been?!'*

I remember one visit I attended – it was a visitation from RAF Cranwell by a lot of 'Top Gun' fighter pilots. They were clever blokes. We ended up in the fermenting room. Above the fermenting tanks we had what's called 'Yorkshire Squares' over the beer. The pilots were asking

'How come, if you say these tanks are sterile, that they're open at the top?'

So I told this pilot that there was a layer of CO_2 – carbon dioxide, a colourless gas which is denser than air - above the surface of the ale. They didn't believe me. This pilot says 'Whoa…come on, now! A layer of CO_2? *Can't* be!' They knew a bit about science. Well, I didn't want to be funny, but he decided to stick his head over the tank to see if I was right so I let him. He soon shot back up after getting a whiff. We had to almost pull him out.

‘He’s right!’ he said, ‘it’s CO_2 alright!’
There are lots of other memories of the many good times staff enjoyed. Mansfield Brewery was an efficient, highly productive business. It is so sad that it’s no more. At my level in the company I didn’t understand the ins and outs of why the Brewery went to W&D, I just feel that it should never have been allowed to happen. You can’t go back; it’s over, and that’s that, but it was a fine company, and a great job.”

❖ ❖ ❖ ❖

Chapter Six:
To The Bitter End

The recollections of Kevin Hall, Senior Maintenance Manager, MBC and Neil Tomlinson, MBC General Manager.

Kevin Hall started work in January 1986 for Mansfield Brewery and worked right through to the take over in 1999. He spent a further four years with the new proprietors, Wolverhampton & Dudley Breweries, and took early retirement in May 2003.

"In 1985 I was working in Sheffield for Tarmac. I saw this job at the brewery advertised in the CHAD newspaper. I thought that it could be quite nice if I got the job, living only five minutes from the Brewery as opposed to driving an hour and a quarter back and forth each day to work. I was a site manager at Tarmac on house improvement projects, but I had no qualms about giving it up. I applied for the job and was interviewed by Steve Price and Hugh Wilson and eventually I was pleased to be offered the post. It was vastly different for me. We'd been flitting from house to house. But when I started work at the brewery, I thought that I'd made a mistake, because for the first five weeks I didn't seem to do anything. Then, literally, the floodgates opened and the work began. It never stopped for eighteen years but I loved it. I was 39 when I got the job. It was only my fourth job. I'd started my working life building schools, fire stations and civic buildings such as libraries and then 11 years later, started work building the Four Seasons Centre in Mansfield. I'm a joiner by trade. But for that first five weeks at the brewery it was dead – I really thought I'd slipped up. I liked to be busy.

My job was as Property Supervisor to run the team which carried out running repairs on and around the plant. It was all fairly local. We had five joiners, two bricklayers and an electrician. They worked directly for the brewery and there was an outside firm, JKD who did work who had a joiner, a plumber and a bricklayer. Work that needed doing was reported to me and I gave the jobs out. At first I had a bit of sorting out to do, personnel-wise. There were some workers who seemed to vanish and hide away somewhere. I tracked them down to a little café off Mansfield Market Place and we soon got things on an even keel. At one time they'd been able to do exactly what they wanted to do because there was very little supervision – so they'd disappear to the café first thing in the morning then later turn up at the works canteen for breakfast, so I had to put this right. Once I'd got the team and the work properly organised it turned out that we'd have four days of the week left with little to do, so we began doing all kinds of work on pubs, first locally and then within a fifty mile radius. I used to do a lot of surveying for the Free Trade team and it came about that they were paying a fortune out for signs made by professional sign makers, and I knew that we could make these just as well. Our department had to make a profit so we had to pay attention to costs and charges. I'm a 'people person' and I like to get involved, and I'd never ask anyone to do something I wasn't capable of doing. Of course there were a few who resisted being 're-moulded' but we soon got it running OK. They were potentially very good craftsmen but if you didn't keep an eye on things they would slip back into their own ways of getting a job hastily finished then leaning on their benches for three days. Most jobs went well but I do remember a bit of a mess when we built a brick bar surround for the beer tasting area. But generally, considering we only got half the time we needed to do a job – because production had to carry on – we did some good work."

"We developed into the free trade side. We began to do building work and this continued right up until the take-over. One thing we began making was picnic tables. The market for these when we started wasn't very good. Instead of pubs paying £85 for these we were turning them out at £39. But soon the orders went up from 125 tables to us making over 1,000 tables in a year. There's still one in my back yard now which has been there over 17 years. This was good because it kept us fully occupied. We used to do work at some of the directors' houses. The MD used to like us to take sawdust all the way

to East Midlands Airport for his horses – but he didn't have any more after I charged him for fuel!"
"Of course we still had the odd problem with the team, one man in particular who'll remain nameless used to go AWOL – he'd never be in the place you'd sent him or expected him to be. Mind you, when he put his mind to it he was a really good joiner, but at other times he'd string a job out as long as he could. My background involved quantity surveying. When a pub was changing hands, I was the person who did what they called the 'dilapidation survey'. I would survey the pub for things like the state of decoration, draw up a schedule for the work which needed to be done before the change-over, or alternatively, we charged them money. When I started there were only around 25 change-overs a year, but soon it went crazy. I used to enjoy this part of the job because it got me out and about. You used to see some real sights. One of the first I did was the Ship Inn at Colwick, which is near Snaith. I'd made arrangements with the old couple there to go in, and early one morning I found the pub. When I arrived the doors were all open and I thought the place had been on fire; black dust everywhere, pouring out of the doors. It turned out that the old girl in the pub was burning stuff in the stove, the old man had been rattling at the embers with a poker and all the soot, etc., from this was floating around. Then out of this gloom trotted this bear-sized Alsatian dog. I thought I needed a change of pants. I had a look around and discovered that the old man was in the ladies toilets – hosing the carpets down. I went upstairs and did the survey and the first room I went into was the kitchen. It was awful. There was this four inch gap between the cooker and the kitchen worktop and I was wondering what this curious stuff was, wedged in the gap and sprouting little bits of wood. It turned out to be solid, rancid fat, and every time they'd lit the stove they'd stuck their match sticks in it. It was revolting. I did a quick survey and on the way back in the car the effect took over and I pulled over, took off my clothes and drove home in my underpants – it had been that bad.

"When we were making signs for the Free Trade, our area sales manager used to ask us get a price for the job and our department would invoice the Free Trade department. Later on in the year, when everything was accounted for, the Free Trade Director, a dear old fellow called Dennis Foster, who is no longer with us, refused to pay. He asked 'Who authorised this… and this?' And it would be John, or Bill or someone. So he said that they couldn't sign anything – 'there's only me who signs for costs in this department.'

So I sent the lads out to one of the sites and got them to take the signs down and bring them back. Within an hour my phone was ringing. Dennis said 'Why have you taken these people's signs down?'

I replied 'Well…. they're mine, because you've refused to sign for them.'

He said 'Well, you'd better go and put them all back up…' So I asked him 'So… if I bring these authorisations up to you, are you going to sign them?' And he did. After that we got on like a house on fire.

"Like the other workers at the Brewery we got our beer allowance. It was two bottles a day but shortly after I started we went onto cans – a tray of 24 cans of beer a month. There was a time when the office staff, back in the 1970s I believe, in their wisdom took money in lieu of beer. It worked out eventually at about ten bob (50p) per week. At the time it was the women in the office who said they didn't want the beer but the money, and the blokes weren't really concentrating, but in some ways they shot themselves in the foot. The lads in production continued to get their allowance, and when you think of it, a beer allowance is inflation proof and 50p per week wasn't. After a while the beer allowance was re-introduced in the office but only at one can per day, because they'd had this ten bob a week. So it wasn't a good idea to take the money instead of the beer. At one time the office staff used to pick up their allowance from the Wines and Spirits department on Great Central Road. Then it became Gordon Kendall's job to distribute the allowances. He used to open his little shop up just for one hour – and if you missed that hour then you'd lost your allowance. Gordon had everyone's allowance there, but if you weren't there on time, then you couldn't collect it tomorrow – it went in Gordon's carrier bag.

Our Christmas parties…. And I'll be careful what we say here…were very good parties. We worked hard and played hard but our department got a bit of a reputation, mostly unfounded, of course. Suffice it to say we had extremely good parties and a lot of fun…"

"I didn't spot the end of Mansfield Brewery coming. There was all this talk for twelve months about a possible take-over, and we knew that if it was to be anyone then it would be Wolverhampton & Dudley. I still had a great faith in the British legal system that it would never be allowed to happen – but it did happen. I was as sad as anyone over this. I think everyone would agree that we enjoyed working at Mansfield Brewery so much that they could have buried us there. Yes, the job had got harder under newer conditions, etc., but the atmosphere was still there. The feeling is still there; it's on the bar on the pumps, it's in the air. We still have the old Christmas parties and they are full every year, even though the company has gone – people flock from Hull because what we had together as workers is still there. It's a great night out which is held a few weeks after Christmas in Mansfield's Civic Centre. You can't knock Wolverhampton & Dudley for being what they are – a profit company. But they don't have that culture which we had – which values people. We had an allegiance with Mansfield Brewery".

General Manager, Properties, Neil Tomlinson spent 20 years in service including two years with Wolverhampton & Dudley Breweries.

"When I commenced employment as a property surveyor with Mansfield Brewery it came at a time of transition at the brewery. We were looking to progress, become more professional, take on more staff. I initially saw the job as a surveyor advertised in the local paper and at first was unsure about applying as my father – Arthur Tomlinson – was actually manager of the Estates Department (as it used to be known) at the time. However, I did apply, but I'm not sure that Dad thought it was a good idea! Needless to say, he didn't interview me.

After three interviews, culminating in a final interview with Properties Director Hugh Wilson and Managing Director John Hings, I was successful in getting the job. When I started the job I did, in fact, have to work for my father and he certainly lived up to his reputation for being a hard taskmaster. In those days any new employee went through a four-week training period known as 'induction'. This was a very good idea and, depending on the type of job you were going to do, you experienced some time in every department and absorbed the workings of the Brewery, the culture and so on. Sometimes I had a 7 a.m. start in the morning to go out on a dray. Even then, my father would expect me to return to the office to complete drawings for Free Trade or other such work – he was determined that I wasn't going to have an easy ride whilst working for him. He did have a bit of a reputation and I had to get used to a 'telling off' (to put it mildly!) every day whether I needed it or not. But I must say that this put me in good stead for dealing with some of the characters I continued to encounter during my years with the company.

My father worked for the company for 28 years, most of the time with Harry Ridley who was wines and spirits manager – and Charles Kerridge who was sales manager. They all reported to Robin Chadburn and these four were all great characters. To an outsider they seemed to run the company and to have a brilliant time whilst doing so. However, things had to move on and the company needed to progress. PLC came and in the early 80s a major influx of new employees at all levels. This was coupled with the new facilities and acquisitions and massive investment in retail outlets and free trade. The company invested in a brand new 'state of the art' brewhouse together with the acquisition of North Country Breweries of Hull. The export business also reached as far away as America and Russia during this time."

"Throughout all this activity and change, the company incredibly still managed to maintain that same camaraderie and commitment that prevailed during the time of my father and his peers. In fact, this same feeling still exists today to a certain extent even though the company has been closed for the past four years. Whenever ex-Mansfield employees meet up together, for whatever reason, there is a continued atmosphere of all having once belonged to one big happy family."

The company continued to expand into the late 80s and right through the 90s, enjoying continued success and building a magnificent new office block along the way. My most successful outlet development as a property manager was undoubtedly the Swan in Mansfield's town centre. In its heyday, the Swan Hotel was a large coaching, banqueting and restaurant operation. We actually demolished half the original property and outbuildings, spending around three quarters of a million pounds, which at that time was a huge sum. The Swan became one of the most successful town centre pubs in the country and still has more than 30 derelict bedrooms on the upper floors.

Lloyds No. I Café Bar outlets were another major concept that I worked on, the first one being opened in Nottingham Lace Market. This project ran into massive problems with planners, neighbours, contractors and lawyers and it seemed as if we'd never complete the project on time. Indeed, I spent the whole of the final night in the pub trying to get the work finished and actually slept on the pub floor that night. We only just made it in time for the magistrates to visit in the morning but Lloyds went on to be a leading Café Bar brand. I was lucky enough to oversee an annual project and maintenance programme that at one stage was the envy of the industry. It has been said that we were profligate in having a fairly large maintenance budget, but I prefer to think that because our estate was in good condition this reflected in the customer experience and satisfaction, motivation of the licensees and ultimately profit and asset value.

As a department the Properties Team were responsible for dealing with everything from bees to bats, to cranky landlords and awkward authorities, but still managed to carry out the jobs with a sense of humour and fun whilst maintaining the required level of professionalism. This was the culture pervading through the company - it was one to cherish and retain. It was a unique culture and unfortunately will never be recreated.

The news of the company's imminent demise came as an utter shock to me. In fact, I was on a plane returning from holiday, having seen no English news for two weeks, and noticed in the newspaper that our shares had risen dramatically. But I speculated that this must mean that we had taken over another company. This was because before I went on holiday I had been looking at a pub estate in Oxfordshire (Morlands), with a view to acquire. This thought, however, was soon dispelled when we arrived back in the UK and got into our airport taxi. The driver was an acquaintance of mine and he informed me that I was now a Wolverhampton & Dudley employee! My reaction was 'Don't be stupid – you've got it wrong!' But he turned out to be right. None of my family back in England had mentioned it in our telephone calls home because they knew how much of a shock it would have been and how it would have had an impact on the holiday. They knew that there would have been nothing I could do anyway.

Back at work it was devastation. I had never seen anything like it. The reaction was 'This can't be happening', as the workforce believed that because the majority of shares were in a charitable trust then the company could not be taken over. It could never happen. The senior managers decided to get together to see if anything could be done to rescue the company and we concluded that we would establish a fighting fund to challenge the Trust's decision in the High Court. There was overwhelming support from employees, customers, suppliers and contractors and we eventually had our day in court. The rest, of course, is history."

"The overriding factor for me is that working for Mansfield Brewery plc was an utterly brilliant experience and working with, and for, some fantastic people was an absolute pleasure, one which I'm sure could never be repeated."

❖ ❖ ❖ ❖

Chapter Seven:
Beauty and The Brew:
Some stories from The Ladies of Littleworth.

Still enjoying the memories: L-R Madge Branfield, Bill Donbavand, Barbara Brown, and (seated) Alice Donbavand and Jean Elliott.

Although, in general, brewing beer seems to have been primarily a male occupation, as with most businesses it has to be the women who keep the wheels of commerce in motion. On a fine late autumn afternoon, a quartet of Mansfield Brewery's finest girls assembled at John Else's house for a nostalgia session – and even allowed one of the husbands along – but only because, like them, he too had worked for Mansfield Brewery. MBC's accounts and administration offices were populated by some very popular and well liked women, all of whom appear to have had as much affection for the company and the same love of their job as the men. In this meeting some of them appeared a little bemused.

Barbara Brown, Madge Branfield, Jean Elliott, Alice Donbavand and her husband Bill combined have all clocked up the best part of a century or more together in the service of the brewery.

Madge has the longest service. She served so long that she can't exactly remember when she started at the brewery;

"Probably mid 1960s. I left school and worked for a while at a plumber's merchants. I saw the job at the brewery advertised in the paper. I seem to remember I was interviewed by the Company Secretary, Fred Gleadell. My job was in company accounts."

She recalls that there were 12 girls involved in secretarial and accounts work. "I used to operate the company accounts machine – it wasn't anything like the computers we have today. It was a big machine called a 'comptometer' It was a kind of book-keeping machine. It was purely a 'men-only' environment in those days – all the main managerial staff were men." It was known at the time that the men in the company received perks such as their beer ration, but as Jean Elliott, who joined the company in the 1970s recalled, "We didn't – but we did get a glass of Oloroso sherry on our birthdays."

(This is humorously disputed in the gathering; some seem to remember getting a bottle rather than a glass – others recall that they were allowed a small bottle of orange or lemonade. Not so much that the demon alcohol would spark off lewd behaviour; more, as Jean says, that "You see…we were considered too innocent.").

Jean continues.

"I worked at Metal Box for five years after leaving school, then went to the Wakefield Building Society. I hated it and promptly left. I was probably about 20. It was a very different life working for the brewery than what I'd experienced at Metal Box or the building society. I was young and I had a good social life. I used to go dancing at the Mansfield Palais and other places. It was during those happy days that I met my husband – and no, he didn't work at the brewery. Our supervisor was a lovely lady called Doris Watson, but our boss was not always the most popular of men. When he was nice, he was very nice. But the rest of the time… oh dear. Not popular. In those days I remember we used to have our annual cocktail party in January at the Swan. It was all silver service then. We started going to the Aquarius. I remember one year they had a swan all carved from ice."

Alice Donbavand started work at the brewery in 1964 and retired in 1991. "Our Christmas parties were quite something. I was in the properties department and we used to hire one of the pub rooms and invite the whole of the brewery. Don't ask me how, but we all squeezed into the Portland Arms or was it The Brown Cow? We had a buffet put on, everyone would come down there and then somehow we'd all go back to the 'official' party at the brewery. Then they stopped us from having our own party at the pub so we eventually ended up at the cellars. Wines and Spirits put those parties on – and they realised that what we wanted was a bit of music and a bit of space. But we did usually manage to have a bit of a dance at the cellars." Barbara remembers one such occasion.

"Yes, I remember having a dance with Mr. Robin. (Robin Chadburn, the Chairman). We did a waltz." Alice also has her 'Mr. Robin' story.

"Yes, I remember the first Christmas dinner I attended at The Swan. It seemed all very low key and everyone was standing around and this attractive man – who I thought was a waiter – came up to me and said 'Would you like a sherry?' And I said 'Oh…yes please…' So he went over to the bar and poured me a sherry and presented me with it and I casually said 'Oh, thanks very much…' And Doris Watson came over and said 'Do you know who that was?' And I said he seemed a particularly good looking barman. 'No', she said, 'No, not a barman - *that's* Mansfield

Brewery...'.

"In many ways I considered working for Mansfield Brewery as being as part of a family. I used to walk past the building and say 'look – that's mine'. There were some marriages among the staff and liaisons during those times, and, as one contributor who wishes to remain anonymous reports, "Yes...and the odd illicit relationship going on". In the main, the girls loved the period they worked there, especially under the Chairmanship of Robin Chadburn. They remain united in their pride in having worked there.

"The real characters," says Alice, "were the 'Three Musketeers' Harry Ridley, Charles Kerridge and Arthur Tomlinson. Mr. Deakin was a gentleman. He'd always tip his hat when passing the ladies. At Christmas he would appear around the building with a book of draw tickets and pass among the girls handing them out very politely..."

Barbara Brown started in 1970 a month before her 20th birthday.

"I was a pen and paper invoice clerk to begin with and then I progressed onto a typewriter. I used to deal with invoices mainly for wines and spirits. I worked for Doris Watson, who was like a Mum to all of us. I still keep in touch."

Jean Elliott also recalls her clerical duties.

"When I became a credit controller, our boss, the chief accountant, had this idea that the publicans and the office staff should meet. He took me along to a large gathering – it was in one of Mansfield's big pubs, I can't remember exactly which. Everything seemed to be going along well, but I was hoping that, as credit controller, there was no-one there whose beer supplies I'd stopped. I looked around and couldn't spot anyone. Then the boss said 'Come over here, I'd like to introduce you to this landlord and his wife.' Sure enough, I'd recently stopped their beer and I was about as popular as a slug. It was dreadful, but they hadn't paid their bill..."

Barbara remembers Pearl Guy, the lady who came round with the tea trolley twice a day. "...But we liked to brew up when we felt like it so we had an 'illegal' kettle which we kept hidden in a cupboard. The chief accountant, who often terrified me by peering into the office through his big, horn-rimmed glasses, seemed unaware of this. The kettle was in the cupboard right behind me where I sat, and one afternoon we decided to make a cup of tea. We'd filled the kettle and switched it on and the cupboard door was shut when the boss came in and, unfortunately, stayed longer than he should have done. As it reached boiling point I could feel this immense heat around my feet as I sat there, with steam swirling about my ankles. Then he said "Well, I'm going to have to go because I can see you're all very busy... in fact Barbara is literally *steaming'*...all I was thinking is 'Thank God it isn't a whistling kettle...'. After four and a half years with Doris, I moved with the wines and spirits department into new, larger premises on Great Central Road, where Les Pearson became my boss, then later David Long. In 1983 I moved back to the main offices as Office Services Administrator."

Madge Branfield listens to all this with wry amusement, but when prodded for her own possible misdemeanours, prefers to let the others jog her memory. Jean obliges.

"I remember how Madge used to deal with those difficult customers who used to come into reception. She used to fix them with 'that look".

Madge confirms the recollection:

"Yes – what used to irritate me more than anything was when you had a load of people in reception and Robin Chadburn used to walk in and say in a quiet voice 'Are you being attended to?'. He used to do it every time and I used to think 'If you say that one more time then I'm going to tell you!' Then he'd wander off slowly up the stairs drumming his fingers on the banister – very annoying. I remember operating that 4-line switchboard – especially at that time when we had the black-outs, up until six at night with a torch and a bit of a candle. That was in the three day week period in the early 70s. But we carried on working. They used to have a commissionaire on the door then – Tom – in full company uniform. When it was cold he would come in and stand behind the desk. He loved a cigarette, and would ask me 'Is there anybody coming?' and I'd keep an eye open and he'd light up, saying 'I'm not standing out there – its too cold!'.

Alice recalls the time that her boss called Madge to his office to tell her about her pay rise. "There was also this routine if we went up to the main office to discuss our pay rise. Madge was called up and when they'd finished I saw her come steaming down the stairs and I thought 'Look out – someone's upset Madge…'. I said 'What's wrong, Madge?' She said 'Do you know what that silly bugger's just done? He called me up there to talk about my rise – pitiful though it was - ' So I interrupted her and said 'So what did you say to him?' Madge replied 'I just looked at him and I said b---s!' (Madge doesn't want us to spell this out, but suffice it to say that the word refers to spherical objects.)

Nobody's fool, Madge knew how to handle the bosses.

"You had to stand your ground with them and have an answer ready or they'd just mow you down."

The directors had a strategy in the wage rise department, as Barbara outlines. "They would tell you what your pay rise was, then say 'but don't tell anyone else'. Well, with this method you didn't know how you stood, if you were right or wrong. You weren't suppose to discuss it."

Alice remembers what a diverse group the directors were.

"There was a sales director – big chap, built like a rugby player, (actually he *was* a rugby player) called Peter Watson. He came into the properties office where I worked with Arthur Tomlinson and said 'Can I use your phone?'. Well, I was on my own in the office. It transpired that he was selling his house. He sat down at the desk and phoned this estate agent and he was trying to handle the deal. The house he was trying to buy belonged to this old lady. But she wouldn't budge on the price, and you could tell that he expected her to – it was 'I'm the brewery' and he supposed he was going to get his own way. He said 'But she's only an old bitch! Get her to change!' The estate agents told him that she wouldn't. He exploded, saying 'Well – she can *stuff* her house!' and with that he slammed the phone down then threw it across the room and it landed in the corner. I wasn't going to pick it up. It was picked up eventually – not by me – when I'd explained what had happened…."

Some of the directors occupied a warmer place in the ladies' hearts, as Alice remembers.

"Mr. Gleadell – Fred, the Company Secretary, was a lovely man, and a gentleman. And Mr. Scruby, of course. They always turned a blind eye when we girls had our little moments of fun – which we had quite often. Mr. Gleadell would often give you a lift home. At one point he lent us a brewery car. I remember when my mum and dad were coming to stay with us from Liverpool, and Mr. Gleadell said 'How are they getting here?' I told him that they were getting a bus to Chesterfield bus station and we were going to meet them there and get a bus from there to Mansfield. He said 'But you haven't got transport, have you?' I said 'No.' He said 'Take my Morris Traveller and have it for the weekend.' This was the Brewery's spare car. Mr. Gleadell said 'Don't make them get another bus – use the car and pick them up at Chesterfield.' Although we did take the car back the following morning, he asked us what we were doing that day and I said that we'd dropped our parents off at Newstead Abbey but he insisted that we keep the car until Monday morning. He was a lovely man, a kind man."

Barbara recalls the pleasant social aspect of brewery life.

"On Friday afternoons, for instance, and I thought this was rather nice, they'd all have a drink together. There'd be a two gallon jug went around, and all the supervisors, the foremen and even Mr. Deakin, the Head Brewer, would come to the Wines and Spirits for a drink. He was an important man – always referred to as 'Mister' Deakin, but he'd join in on Friday afternoon, even though Christian names were never used."

Bill Donbavand remembers the somewhat comical transportation of the beer for these occasions.

"We used to come down from the Properties department to get the beer. I'd travel with Len on his Lambretta scooter with two jugs. We'd be going – rather slowly – up Brunts Street and there was always a Bobby stood at the top. He always knew that ten minutes later we'd be having a drink."

Jean remarks that drink did on occasions see people worse for wear.

"We had a shop at one time on Broadway in Mansfield and our Wines and Spirits department were promoting some product from Australia. We had this huge inflatable rubber crocodile at the side of the counter. Well, I won't name names, but one bloke came in who'd had a bit too much and he decided that he'd be Crocodile Dundee. He started wrestling the crocodile. He was all over the floor, struggling with this thing. Customers were coming in and were taken aback by this hilarious sight."

Alice has similar memories of a more relaxed time:

"Of course, the breathalyser wasn't as prominent then. Arthur Tomlinson and Harry Ridley would go out together in the afternoons. Arthur's job was, of course, to visit all the pubs, so you could say that he and Harry were more 'mellow' after lunch. Arthur was a good ambassador for the brewery and was highly regarded. Everyone respected him. When I was in the old brewery offices I was by the window which overlooked Brunts Street and I could see Arthur's car when it would move erratically down the street. I had to know where Arthur was usually – or at least have a rough idea where he'd gone – in case anyone needed him. One afternoon about three o'clock Robin Chadburn called me and said 'Is Arthur in?' I replied 'No…I'm afraid he's not…' so he said 'Well, when he comes in would you tell him I'd like him to come up and see me?' I said 'I don't think he'll be coming back because he's got quite a lot of calls on today so I don't think he'll be coming back'. But Mr. Robin said 'Yes, but if he does come in, tell him to come up and see me.' I was a bit concerned so I knew some of the places Arthur might be at and started calling around. Somehow I had to stop him coming back to the brewery. Every place I called they were saying 'No, sorry, he's not here…' but I was desperate, saying 'Look – don't pretend – if he's there, tell me, I know what he's like – if he's told you to tell us he's not there, forget it – I need to talk to him…'. But the replies were all the same 'No, he's not here…'

Just as I'd dreaded, I'm sat at the window and at five minutes to five Arthur's car started snaking down Brunts Street. I thought that if I could get out of my office quickly and down the stairs I might be able to stop Arthur. Unfortunately, Robin Chadburn's window was beneath mine and he'd seen Arthur's car too. When I reached the stairs I saw Robin standing there. There was nothing I could do. Arthur came in and shambled up the stairs – it had been another of 'those' afternoons. Robin saw him and said 'Arthur – can you come up and have a word with me?' Arthur replied amiably, 'Yes…of course…' but actually he didn't know at that point what anyone was talking about. He had this phrase when he was like that; he used to reply to everything by saying 'Yes, well – this is it. This is it…' So I'm stood outside Mr. Chadburn's door listening in and all I can hear is Arthur saying 'This is it..'.

"The next morning Arthur came into the office and was sitting at his desk drumming his fingers. I said 'what's the matter?' Then he said 'I came back to the brewery last night, didn't I?'

I said 'You certainly *did*….'

He looked a little puzzled. 'Oh…what happened?'

'You had a meeting with Mr. Robin…'

He was even more confused. 'What was it about then? Did you hear what I was saying?' I told him that all I'd heard was him saying 'This is it'. He obviously didn't remember a thing about the meeting. Those were certainly the pre-breathalyser days – and Arthur never did know what that meeting was about or what the consequences were. But I worked with him for years and Arthur was very good at his job. He had a good sense of humour, and he was a valued member of staff and a nice man. Sometimes our working practices might come under scrutiny, such as when they brought in a Work Study expert – what you'd call maybe time and motion today – a man called O'Donnell. Big chap. He went around every department, changed every system – and made a complete mess of everything. He carried out one of his projects on the deliveries and the very first morning he messed it all up. Edward Deakin was stood there observing all this – and expecting it to happen. He was very quiet, let it all go on until there was complete chaos and then said 'Stop! Back to the old systems – now!' One morning this Mr. O'Donnell came into the properties office and Arthur Tomlinson was there. O'Donnell was a big man and Arthur was quite small, but Arthur always carried a tape measure in his pocket – he used to be an undertaker. He got up, took out his tape and started measuring this work study man up, murmuring 'Mmm…six foot six,…. six foot seven…' We never saw Mr. O'Donnell again."

Prior to joining Mansfield Brewery, Bill Donbavand spent his earlier working life as a ship-builder in Liverpool. He had worked on such legendary vessels as the Queen Mary and Queen Elizabeth, as well as many Royal Navy ships.

"National Service was still in force then so I went into the forces. With the lack of work in Liverpool then, after my service I was looking for work and had an uncle who worked at Brackenbury Builders. I worked with him for a while, then he told me there was a job going in the joiner's shop at the Brewery, so I applied and started in 1965. One of our jobs was sweeping the chimneys in the alms houses on Nottingham Road."

Alice fills in on this part of Bill's story:

"Robin Chadburn was the trustee of the alms houses. When any of the old ladies who lived there died, someone from our joiners' shop used to go in to the houses, do any carpentry which needed doing and paint the houses up for the next old lady who would be moving in. The only thing which had been difficult to get done was the chimneys. Bill and I lived in this bed-sit flat and we'd bought a set of sweep's brushes – we used to sweep our own chimneys. So we very foolishly said one day that we could sweep the chimneys. They offered us ten shillings a time, and as we were saving up to buy our bungalow at the time, it seemed worth taking up."

"We'd been to one of the alms houses," says Bill, "and we'd got the

bag of soot from the chimney and there was a wedding at the church at the bottom of the road – the one opposite the Imperial Laundry. When we appeared on our push bikes the wedding party thought they'd ordered a chimney sweep! We used to sweep chimneys at weekends or at night. I was only on ten pounds a week so that ten bob was quite handy. I often had some odd jobs to do. I remember one event which involved a shotgun. The brewery owned a banqueting establishment in Barnsley called Keresforth Hall. I came back to the office one day and Arthur said 'Mr. Chadburn wants to see you…what have you done?' I said 'Nothing that I know of…' So off I went to Robin's office. He invited me in and said "We've got a problem in Barnsley at Keresforth Hall. The local farmer has blocked the sewage outlet off." The drainage from the Hall ran down through the farmer's yard and had its own mini sewage outlet. The farmer had fallen out with the manager at the Hall and had stuffed tarmac and old sacking up our drain outlet so that when all the sewage was being discharged from the pub it had begun appearing in all the wrong places. Mr. Chadburn said 'I'd like you to go up there and sort this farmer out…'. So I took one of the lads with me and told him to follow me up there. I told him to stay in the pub and not come down to the farm unless I called him. So off I went to the farmyard and there was this broad-Barnsley farmer stood by the outlet with a shotgun. I looked at him.
'That's not for me, is it?' I asked, looking at the gun. He pondered for a minute and said 'Nay, lad. Tha's got a job t'do so tha'd better get on with it…' So I got down into this manhole and got a rope tied around all the stuff that was blocking the drain and said to him 'Just get over there and pull this rope.' So he did and then '*Whoosh!*' - all this stuff – the full sewage menu from the pub surged out into the farmyard. Keresforth had to be re-built later after it had caught fire. I remember being called out one Sunday when the fire happened.
There was another shotgun incident at the Little John pub at Ravenshead. The tenant that day was at the races and whilst he was out the place had caught fire. The gas burner in the bathroom had caused the blaze which had gone out through the roof. I got there and

the Chief Fire Officer, who I knew, said to me 'Bill – I need you inside. We've got a problem here. There's two shotguns – with ammunition – in the building. We think they're on the landing.' And there was another problem. The pub had one of these giant whisky bottles in which they'd collected all this money – quite a sum. They managed to put the fire out and I found the shot guns and ammunition and put these in the boot of my car. We couldn't leave the big bottle full of cash. So that went into the boot. Driving back to Mansfield I suddenly thought – what's it going to look like of I get stopped? Two shot guns and a great big bottle stuffed with cash? Anyway, when the landlord got back to his fire-stricken pub he called me up.

'What a mess, Bill. And my shotguns have gone and our lass's big money bottle – that was her savings for our holiday – that's gone as well.' I didn't let him know anything at that point.

I said 'I'll see you in about half an hour.'

When I arrived at the Little John he came outside and I opened the boot and said 'Get this lot out of my car!'. He was rather grateful. Another problem occurred at the Jesters in Mexborough – it was a night club that we'd opened in tandem with the Aquarius in Chesterfield. The toilets had become flooded. When I got there, there was a fire engine around the back pumping the water out. I went into the car park with the landlord, Ian Gittings, to find the man-hole. We found it, got the lid off and it was 25 feet deep, so I went down, and discovered that the drain was all blocked up with stones. So I started removing the stones and some bloke came and parked his car over the manhole just as about three tons of sewage broke loose. When I finally got out of there Ian said 'I think you'd better get changed, hadn't you?'

This had been a late night call out and it had been dark so the bloke who'd parked over the man-hole hadn't realised it was open. I'll bet he needed a new sump because that sewage came out with some force – and you could say that that was a job where I was 'dropped right in it'. I think Alice was a bit puzzled because I arrived home in different clothes than I'd gone out in.

"When the brewery took over Hull Brewery I had to go out on surveys to their pubs. One was in Beverley, the King's Arms. I stayed in the Lairgate Hotel because the King's Arms was full. I came back one night and couldn't get in my room – then they told me they'd moved me to another room. Working for Hull Brewery was interesting. I wasn't familiar with Hull before that but coming from Liverpool, with Hull being a port, there was that same feel in the pubs and with the people there. There was an odd feature in the brewery's offices on Sylvester Street – the creaking chimney."

Alice and Barbara also remember this phenomenon. The tall brewery chimney had offices built around its base.

"When I was sent to Hull for a while," says Alice, "I was working in the office and one of the chaps there, Arnold, popped in and said 'Do you want anything, Alice?' and I said 'Yes – I want to know what that noise is…' He said 'Oh, that's the chimney moving…' So I said 'I think I'd like to sit somewhere else!'".

When Hull Brewery was closing down and all the furniture was to be sold off, Barbara was sent there to do the inventory.

"I was with a chap called Norman Smallfield and we were writing lists which included things like 'chair, one, leg missing' and 'pedestal desk…pedestal broken' stuff like that. When Norman came in I said to him 'Is there a ghost in this brewery? I feel like I'm followed by somebody in every office I go into…' He said 'Ah….no – that's the chimney.' When I put my hand on the chimney wall I could actually feel it moving. It was a bit creepy, to say the least…"

Bill has memories of another Mansfield Brewery pub, the old Wagon & Coals, which had to be demolished to make way for new offices for the Electricity Board. "When you looked at the pub from the outside it seemed odd because it was slightly bigger than the space inside. To begin with, we found a Priest Hole. We removed some panelling and there it was. The pulley was still there which they used to ring a bell to alert the Priest. I remember walking up the stairs and getting to the landing and thinking 'Funny – I ought to be turning left here yet there's this wall, papered over…'. So we went behind the wallpaper and there was a wooden, tongued and grooved wall. Behind it we found a door.

I opened it up and in this secret room there were two *chaise-longues* – a ruby-coloured one and a blue one – and we couldn't touch them because the chap who'd bought the place to demolish it had first call on everything other than we'd specified. But we found no end of old coins in there and we removed some lovely old timber beams which we used for another job."

Another unusual feature at Mansfield was the brewery's bottling tunnel, where bottles were moved along by conveyor belt. Barbara remembers the tricks often played on the girls should they have to pass through the tunnel. "They'd see us go down there," says Barbara, "and you'd get half way through that wretched tunnel and someone would switch the lights out so you were in complete darkness. They'd often do this. We used to run down the tunnel because we knew they'd pull that stunt every time." John Else recalls other opportunities the tunnel sometimes offered.

"There'd be two dozen bottles start off at one end and only half a dozen arrive at the other – so there'll be some blokes emerge with very lumpy pockets…"

With the exception of Madge, who retired in 1987 after 37 years, and Alice, who had served 27 years and left in 1991, the others saw out their time until the end, but there was a feeling among the girls in the office that the social atmosphere had changed as the new board members replaced the old one by one. The camaraderie among the men brewing and bottling may have continued apace, but closer to the seat of power the girls felt that there was a less caring atmosphere developing. It is a mark of their confidence in their work and their company that they felt they could respond to this in such a direct way, as Alice remembers;

"So many of the girls were commenting to me that we didn't see anyone much and that the nice little social touches we'd once enjoyed seemed to have gone. It was my 25th anniversary and I came in from lunch one day and there on my desk was a letter. It said something to the effect that 'you've got 25 years of service in with the company. Here's a catalogue, choose something up to £131 and don't forget that I'm the one who has to go and pick up your choice and I have to do this in a certain time…etc. etc.'

"Frankly, it wasn't a very nice letter. It made me feel that by having 25 years service in I was an inconvenience. Now I'd never felt like that with the company before, because they'd always made me feel good. So I sat down and I wrote to the board. It was a simple letter – I knew there was a board meeting coming up – which informed them that we all thought that they'd become very remote, and that there was a widening chasm between staff and management and unless they did something about it - get up out of your chairs, leave your offices and start meeting your staff and talking to us, then you'll lose a good proportion of your good staff.' I sent the letter off but Eileen Ashley, the directors' secretary, copied it to every member of the board. The director of properties who was my boss came down to see me. Board meetings usually lasted about an hour but the arrival of my letter had led to a two and a half hour board meeting. I didn't know when he came down that everyone had had a copy. He put his head in our office and said 'It's hit the fan today…' What made the directors more remote was that at this particular time we weren't in the main brewery building but up at Dame Flogan Street in Mansfield. I was told that Ron Kirk, one of the bosses, was coming to see me. I said 'Does he know the way?' But when he arrived it was a chance to clear the air and explain things better. Maybe it didn't change a lot but it made them take notice…."

But, despite the temporary remoteness of higher management, social activities still featured in Brewery life. Barbara Brown recalled how the Mansfield Brewery Walking Group started.

"It was started by Gordon Henderson, who worked in the boiler house, and his wife Margaret. They were both very keen walkers until Gordon had a bad accident, he was knocked off his bike so was out of action for a while. Clive Almond who worked in the properties department stepped in as walks' organiser with his wife Molly. He used to take us on some lovely walks on Sundays. We'd walk in all kinds of places, such as Derbyshire, country parks etc., and this really took off because at one time as many as 55 employees turned up for a walk. On some evenings in the Summer months we would walk until dusk. The walking group has continued despite the closure of the Brewery and its members meet on the second Sunday of each month. An annual dinner is held at the Robin Hood Inn at Baslow, preceded by an invigorating walk through Chatsworth Country Park. Also each year a weekend away is organised."

"At Christmas time, the Baily Thomas Provident Fund provide a hamper and a turkey for those brewery pensioners who took their pensions when they left, (as opposed to those who left and didn't take up their pensions)". Alice also recalls these Yuletide treats:

"There were an amazing number of chickens given out by the brewery each year, usually from the back room at the Portland Arms. On Christmas Eve morning, Ernest Fletcher, a lovely old bloke, a labourer in the workshops, used to sit with a bucket of hot water drawing and plucking these chickens." And later, as Barbara recalls, there were also frozen turkeys, the distribution of which she, along with her assistant, Vicky, was involved with.

"There seemed to be thousands of them. They'd arrive very early in the morning, about seven o'clock. The first job was to sort out the managed house allocations. All the managers got one in the pubs. Originally they got two chickens and then it was changed to one turkey. Our area managers used to have to go out to the pubs and deliver the turkeys, so we'd have all the boxes stacked with the relevant area managers' names on them. It was great fun but it was damned hard work and our backs used to be killing us by the end of the day. My brother, David Starkey, worked for the brewery with managed houses as a stock-taker. He was a member of the infamous Whippets and Ferrets." One of the main Whippet and Ferreters, John Else, remembers how David entered into the spirit of their bizarre tours.

"Yes, I remember we were in Wainfleet All Saints and David had a tot of everything mixed in one glass. His nose – I can see it now – went a bright red…later I thought he had died."

Jean looks back on those sad days at the end when her and her colleagues all knew that their jobs were coming to a close.

"We'd been told that we had severance pay and we had to go to the brewery to pick it up. There'd been a period of ten days' notice and at first it looked as if the shop would remain open, but John Bartle told us that even the shop was finishing. I went down to get this cheque and they kept apologising for the lack of notice – ten days wasn't much to go and look for another job. This girl was sat there handing the cheques out and said 'Mrs. Elliot, here's your cheque…I am so sorry' but I said 'No – *don't* apologise!'

Monday July 31st 2000: A presentation to Barbara from office services and reception staff on their final day of employment.
L-R Isabell Cooling, Rachel Hallam, Emma Morris, Vicky Phillips, Barbara Brown, Carol Brooks, Pauline Robinson (cross legged, Linzi Whitemore).

I suppose that was my little tantrum. Walking out of the building for the last time after thirty years was a real tear-jerking moment. Everything we'd enjoyed there was gone and over. I still can't bear to walk past the building today. For years it had felt like part of your home. We had so much social contact outside the office. We had so much fun."

When Alice Donbavand started at the brewery she still had a strong Liverpool accent, and remembers how, if she wanted something in the office such as a book, they would hold it back.

"They'd say 'go on then – ask for it – what is it that you want?' and I'd say 'pass me that *buke*'

Early 2000: Mike Cleator: One of many snaps taken by Vicky Phillips as the sun set on the Brewery.

and this became part of the office entertainment. I never took offence because it was all part of the fun we used to have…."

❖ ❖ ❖ ❖

Chapter Eight:
The Driving Seat
The Directors

*'I had a taste of the beer when I arrived and found it agreeable.
I've always enjoyed draught beer'.*

Tony Scruby, Managing Director

The board of Mansfield Brewery plc changed over the years to accommodate a variety of helmsmen, some, like Robin Chadburn, more popular than others. One of the most memorable Managing Directors of the company was undoubtedly Tony Scruby. Mr. Scruby today lives in elegant country retirement on his tranquil farm deep in the Nottinghamshire countryside. His memories of life in the Brewery are a good two decades behind him. He came from a different breed of captains of industry; the old school, the gentleman with a hint of military bearing, someone to whom the staff could look up.

"I was an unusual bird in as much as I had qualified before the war as a solicitor. Of course at the end of the war, after five years in uniform, I had no legal experience because I hadn't been able to practice. I had left school at sixteen and was sent to London to read a course before Articles at the Law Society. I was then an Articled Clerk with Godson & Co. in Sleaford. I joined the Territorial Army, the Lincolnshire Regiment as a 2nd Lieutenant. When the war came I saw action in Norway. Whilst I was in Norway the Brigade Commander was a smoker and a bridge player. He hadn't managed to assemble a bridge four, and was short on cigarettes. Well, it went around that I didn't smoke so in return for him having my ration I was invited to play bridge. That really put me at the foot of the ladder for staff appointments. Later on, there was a shortage with a Division in Iceland. They'd sent two officers to the Staff College and one of them had broken his leg so they sent for me. That job really had a big influence on my war. After Staff College I was sent to West Africa where we were to produce a Division to fight in Burma. I became Brigade Major with one of the Infantry Brigades and was in Burma until 1944.

"After five years of war in the Infantry, I was demobilised as a Lieutenant Colonel, but of course, I'd had no practice in the Law. I was, however, lucky enough to be offered a job in the City of London with one of the leading firms specialising in Company Law. I got married – to a Lincolnshire lass – in 1947. I realised that in no way did I want to spend my life in the City of London."

"I was asked to become a partner in an established firm of Solicitors in Nairobi, Kenya, where I built up a large practice acting for the subsidiary companies of international (U.K. and U.S.A.) businesses. Many of them asked me to act as a non-executive director of a wide range of businesses. After approximately six years I was asked to become Managing Director of the International Trading Group's East and Central Africa activities, with a view, in due course, to joining the parent company board in the U.K. As we had by this time three children whom we wished to educate in the U.K., this approach attracted me."

"Back in England I was working for the same International company yet, without those 5,000 miles between us, I realised that I didn't like them much and I think they liked me even less. So eventually I was asked to join an old-established family firm of Solicitors in Essex. I was just about to do so when one evening the phone rang. I was asked if I would like to come up and talk to the Chairman of Mansfield Brewery. He owned a farm close to my in-laws so I think that's how the contact came about, but I'm not sure. I had been a non-executive director of Allsop Breweries in Africa, but I knew nothing about brewing beer. So perhaps they thought I was suitable material for Mansfield. When I arrived, I got a friendly reception but not from everyone – there were murmurs such as 'What the hell does he know about brewing?' But we soon got used to each other. By this time I was 48 and I could see what a successful company this was, and indeed a happy one.

However, I thought the area around Mansfield, much of which was made up of coal mines, was not everyone's choice. I remember that just prior to my starting at the Brewery we had had our annual month's holiday in Scotland. I had gone to a lot of trouble to devise a route for my wife to come back with me which avoided any coal-mining areas. I had this all plotted out but soon took a wrong route and it was one monstrosity after another.
"I got on well with the Chairman and of course Philip Baker and others. In those days there was a kind of division; the brewing staff were over on that side of the site and we the management were on the other. Although the brewing staff would come over to see us, it wasn't done for us to go over and see them.
I had a taste of the beer when I arrived and found it agreeable – I've always enjoyed draught beer. I took over from a man called Jenkins who had started at the brewery as Secretary and eventually became a director. I do recall that the relationship between the Chairman and him was not all that cosy."

Philip Baker's initiation into brewing began before World War II. "I began as a pupil at the brewery in December 1938 under Sydney Deakin. The day war broke out I volunteered for the army – not, I assure you, from a sense of patriotism, but if you actually volunteered to join you could choose your unit. I didn't want to end up in the infantry if I could help it, so I went into the artillery. I went to North Africa, went up and down the desert several times and right through to Tunisia then Sicily and all the way up Italy. I was lucky in Italy because in December 1944 they had a scheme where those who had been overseas four years or more had a chance to go home for six weeks. It was done by putting names in a hat and I was one of the lucky names who was drawn out, so I managed to get home. After six weeks it seemed an utter waste of time and money but they sent me all the way back to Italy. I landed in Naples and by this time my unit had broken up – so they put me in the infantry. I was a sergeant. They sent me for training. Infantry training seemed to comprise nothing more than us running up and down mountain sides, which I didn't find all that amusing. I wasn't very fond of Regimental Sergeant Majors and when I arrived back from England this RSM said 'What the hell are you doing here?' I replied that this was where I'd been sent. He barked back 'Well I don't want you so you can bugger off!' So I politely asked where I should 'bugger off' to. He suggested a reinforcement camp at Forli in Northern Italy. I asked him how I would get there, and he said 'You can find your own way!'. So I caught a train, sitting most of the time with the engine driver on the foot plate. It took a whole week to get there, and when I did no-one had any idea about me at all. So I hung about and was then moved to another reinforcement camp at Rimini. I stayed there doing nothing until I would eventually get home at the end of May 1945. I was moved between regiments and hoped that I wouldn't be sent back to the front because if that happened you lost your chance of being demobilised and sent home. But I did get home and arrived just after the war ended; it was June 1945. In Canterbury I got the job of being in charge of the new Sergeants' Mess – it was a doddle, and from there I went back into brewing. I was lucky because all through the war, Edward Deakin, Sydney Deakin's son, wrote to me every month. It was remarkable really, looking back, because I'd only been at the brewery for a short time. A lot of the letters I didn't get, of course, but those I did receive had kept me in touch and up to date so coming back wasn't so much of a strain. Edward Deakin was really good to me. When I was asked to go on the brewery staff on January 1st 1947, Edward's father retired so Edward became chief brewer. At some time in the 1950s when Edward was elevated to the Board, I became head brewer. In 1957 Michael Parsons joined us as second brewer. There were some great characters in the brewery then. Those men were the salt of the earth. I got a taste for sherry when I got back into brewing. I used to like a glass in the morning. They would frequently send a bottle of sherry down to the brewing office. I remember when they were going through the accounts and the subject of sherry came up with the question 'All this sherry – look at it – all on expenses – where's it all going to?' I just sat there and said nothing! I knew where it was going to! I remember one of the girls who used to come into the office – I think it was Jean Elliott – when the sherry was around and she'd say 'Oh. I like a glass of sherry…' and I said 'You can have a glass of sherry when you come in here to tell me you're having a baby.' I was still saying that to her at my retirement party! Yes, that sherry was all down to 'allowances'. I remember Edward Deakin used to sometimes say 'Let's go down to wines and spirits for a drink'. In those days the wines and spirits department was under the offices so we used to go down into the bowels of this place and crack open a bottle of Champagne. Les Pearson was the foreman in charge down there and every time Les would say to Edward Deakin 'What shall I charge this Champagne to?' and Edward would always reply 'Oh…put it down to 'breakages'!' Les was a very nice man and good at his job."

Philip recalls the formality of board room relations in the early days of his employment.

"Jenkins was not an easy man to get on with. If fact I thought he was almost impossible to get on with. He was not an easy man and was one of the few men in the brewery who would simply address you as 'Scruby' or 'Baker' or whatever – never ever by your Christian name, a fact which didn't go down too well. …"

Mr. Scruby continues:

"You must understand that this was a very successful business employing top quality men and ladies, and my job upon arriving was to keep quiet for a year. So that's what I tried to do. But when I did start to look into the company's affairs I realised that quite a few changes would be necessary. For example, we had a retail operation selling through the side-door of the brewery. I had to ask what our function was – to brew beer at wholesale prices or to be a retailer through the brewery's side door. So I organised a circular which was sent out to all those people who had supported us through our side door to say that this service would be ending and could they please go to one of our outlets for their requirements. In many ways that side door arrangement seemed to be a method by which the brewing staff

could come in off the street to speak to us – as long as we didn't go to their side and bother them! But the Secretary expressed opposition to my circular going out. 'You can't send this out, Mr. Scruby – it could offend some important people.' So I asked who these people might be. Someone was named, so I asked to see his trade card. This was a record of his business, written out manually of course in those days. Well, I saw from this that this particular person who I was about to offend so much, his order that year had been for five cases of Double Dimple Whisky. This was rationed – the company only had six cases. So I said that we should still send out the circular but with a pleasant covering letter. His response – also by letter – was furious. I remember at the next annual general meeting I thought that my first job was to make this gentleman happy. So I sat next to him at lunch and eventually we resolved our differences. He said to me 'You know, Mr. Scruby, you're not as bad a fellow as we thought, so I'm going to do you a big favour.' I replied 'Well, that's very kind – what is it?' He said 'I'm the oldest life member of Hollingwell Golf Club and I'm going to propose you for membership.'
I wasn't even living in the area yet so it didn't mean much then, but I did become a member and played at Hollingwell five times over a three year period."

Philip Baker remembers this as an age before fashions for various beers flooded the market.
"In those days our aim was to keep brewing simple. We made a bitter, a mild and a strong ale and that was it. Just three beers. We went from cask to bright which meant tanker loads of beer instead of it being in casks. That wasn't the choice of the brewers but a response to the choice of our biggest customers – the Miners' Welfare clubs – they wanted beer by tanker. It was successful but to my mind it wasn't proper brewing. It was filtered beer but we had to do it. Mind you, today the circle has turned again and cask beer has come back into its own. Bulk beer was a hugely expensive operation to start with. You had to buy new tanks, tanker lorries, etc., but it did make the

company profit and the miners' welfare clubs lapped it up. But they could be fickle – if, for instance, during the brewing process we had fining problems, the miners would simply go to another brewery for their beer. I recall the steward at Ollerton Miners' Welfare Club – a very good steward who used to order around 30 hogsheads of beer per week – he phoned me up one day and said 'Hey up – what've y'been doing with the ale? It's like bloody marmalade!'

But that sort of thing – fining problems, prompted the big outlets to demand filtered beer. Of course, without having to lug about those great big 54 gallon hogsheads, a lot of labour was saved with tankers – just a pipe down to the club's cellar and fill the tank – but to my mind that was never 'proper' brewing."

Mr. Scruby remembers the capacity of beer consumption when the pits were open.

"We had a pub close by the pit at Ollerton – The Plough. Prior to the shifts changing at the colliery, there was always one hundred pints of beer lined up on the bar ready for the men coming off shift. By the time I left the brewery, over 50% of our beer was delivered by tanker. Today there's none. There is, of course, some filtered beer around, but the days of delivering by tanker have passed. At that time I saw our job as directors as supporting our sales staff. Our beer was mainly sold in bulk to miners' welfares, working men's clubs and other pockets of industry. So that meant that, even as Managing Director, I would spend many nights out there visiting these places. Often I wouldn't arrive home until after 11.30 at night. And of course I visited as many of our pubs as possible – often had lunch in one of them. I would visit at least 50% of our pubs in a year."

Philip Baker also recalls the characters in the company who stood out as sales staff. "There was Austin O'Donoghue – he was quite a character, Charles Kerridge, sales manager, and of course John Bartle. There were others – one whose name escapes me now (probably John Bennett) but I recall that he used to come up to my office and voice his complaints about the beer in several places. He'd make his point rather forcefully and then go. Mind you, that was his job. He was right to do so."

Sadly, as in most industries, the age of the great 'gentlemen' directors is long passed. The 21st century is the age of the computer, the spreadsheet and the accountant. As we finished off our enjoyable lunch in the tranquillity of Mr. Scruby's comfortable dining room, with the spring sunshine streaming through the windows, we were able, through a final vignette from these two elder statesmen from an age where people really mattered as much as profits, to conjure up a lasting vision of a different, almost Dickensian business environment.

Tony Scruby commences the illustration.

"Talking about sherry….. the previous chairman, Claude Chadburn, who lived in the Hall here at Papplewick, was at the time I recall company President. On Wednesdays he would come here and go around his woodlands but on other days he would come to the brewery. At 11 o'clock sharp each morning there would be a rap on my door and no matter what I was doing – even a Board meeting – I had to stop everything. We would go into my office and Claude would have his glass of sherry and then I could get on with the rest of my day. He was an extremely nice man – not so much a businessman but one of the true gentlemen of his age."

Philip Baker agrees. "I recall that once I'd gone onto the staff at the brewery Claude Chadburn was always the one who came to tell you about any salary increase which was planned. This one particular day he came to me and said 'Ah…Philip. We're going to give you a £50 bonus.' I expressed my gratitude, but he touched me on the arm and said 'Mind you, I don't know *what* you would have spent it on…' Two or three years later I said to Edward Deakin

'You know, over two years have passed and I've never had a salary increase.' Edward was astonished and said that I'd better go and see Claude Chadburn. So I plucked up the courage to go and see him in the tiny little office he had downstairs. When he was surprised at something he had this odd habit – a sharp intake of breath – and when I told him I'd not had a salary increase for so long he drew in his breath and said 'Oh…dear… I'm terribly sorry about that, Philip… *I completely forgot!'* He was a dear old chap and I was very fond of him. And I got my salary increase – more than I was expecting, too."

John Else recalls that when he was interviewed for his job as Training Manager by Tony Scruby, Philip Baker, Tony Robinson and Patrick Helps (of Hardy's & Hansons), Claude Chadburn was present in the room – and he slept through the whole process.
"Claude could be a very difficult man," recalls Philip Baker. " To get him to lay out for capital expenditure. You had to work hard to convince him at the start – but he came up trumps in the end. A lovely old man."

Our conversation ended with the subject of the men who have been referred to by some as the 'Gods' of the business – the brewers.
"They were the vital link in the whole company," says Mr. Scruby, "because the ultimate taste of a glass of good beer was down to them.."
With a characteristic modesty which runs like a thread throughout our meetings with ex brewery staff, Philip Baker remarks "Yes… I suppose we were important. But we were no more important than any other member of staff. At Mansfield Brewery we were all part of the same team…"
Sadly, during the completion of this book, Philip Baker passed away after a short illness.

❖ ❖ ❖ ❖

Mansfield's men behind the brew: L-R Jock Nangle, Philip Baker, Robin Chadburn and Managing Director Tony Scruby beneath the beady eye of Claude Chadburn.

1992: Bob O'Meara receives his long service award from Richard Chadburn for 15 years service.

Chapter Nine

When The Circus Came To Town

"We were sat in this hotel and I said
'Derek... I've jut seen a pink elephant walk past...'
Derek looked me up and down and said
'I'm going to get you to bed...'
But then a purple giraffe walked past."

A Conversation between
Bob O'Meara, General Manager, Take Home Sales
and Terry Johnson, Free Trade Regional Manager.

SALES: 'JUST GO OUT – AND DO IT!'

Terry: "I remember my induction in the summer of 1977. The Sales Manager Harry Battersby employed me and his boss was Sales Director Richard Umbers. We had a Regional Manager, Mike Lewis, who reported to Harry. That induction period was fantastic. You did everything. You went out with the draymen, and it was all about the tradition of the Brewery. I remember meeting a chap called Sid in the brewhouse. He did something with the coppers in the brewhouse. His father had worked there before him. He had to tip these huge sacks of hops in – they were taller than me, and I asked him 'How do you know how many to put in?' and he replied 'My dad told me'.

I thought that was fantastic – this was what it was all about – I knew I was in the right place and only wished I'd been there from being 15 years old. I'd started work in accountancy in a steelworks in Sheffield, and when I got the job at Mansfield Brewery I was a salesman, selling cigarettes for a wholesaler in Retford, selling to pubs and clubs. I remember being interviewed by Richard Umbers. I'm from Worksop, and there was one pub there, The Innings, selling Mansfield beers.

He said 'Now, Terry, I want an honest answer from you – have you tried the beer in the Innings?' I told him I had. He said 'What's it like?' I said 'It's awful. I can't drink it.' He said 'I'm pleased you've said that – we've had a few problems with the beer over there – I think we're sorted but will you go and try another one tonight and tell me what you think?' Yet that was the general feeling among the people just 17 miles up the road in Worksop – they didn't like Mansfield beer. But not long after that, several factors came into play and over the next five years everywhere in Worksop started selling Mansfield beers. It's now the most popular beer. When I started, Mike Lewis gave me a set of record cards and said 'These are your customers. Go out and do it.' I said 'Do what?' he said 'Sell beer. Give me a shout if you need me'. I had no experience selling beer so I just had to get on with it. But the customers were great. I needn't have worried. I remember driving for the first time into Rotherham and getting lost. But going from customer to customer – they were so helpful – I began to realise what a good brewery it was. The first batch of brewery people I got to know were the draymen. They were the salt of the earth. Three or four times a week I'd get a phone call from them feeding me intelligence about all sorts of stuff happening at pubs, or if they'd noticed a foreign beer in the cellar of a pub where we had 100% trade – they really kept me up to date with everything. I spent some time with Harry Ridley, the Wines and Spirits Manager. There are lots of stories about Harry. There was one occasion where he'd visited a golf club and it started raining so he couldn't play, so at about mid-day he started drinking. He was still there at about midnight – just him and the steward were left. The steward locked up and they both went to their cars. Theirs were the only two cars left in the car park – and they reversed into each other!"

THE MAN FROM MANDORA

Bob: "The West Midlands was my home territory. I worked for the soft drinks company, Mandora. But prior to that I worked in a slaughterhouse. I was a gland extractor. I used to remove glands from the carcasses; the hyperthalmus, the super-renal, the pituitary and the pancreas. They were sold to Boots for medical research and the pancreas became the insulin for diabetics. I used to put 'gland extractor' down on my CV when applying for jobs but it never landed me one. At the time the Brewery owned a soft drinks company called Eclipse but they had shares in Mandora. There are various theories as to how Mandora got its name. A man called R.L. Jones thought of the name, apparently, but he was killed in a car crash so nobody really knows, but they were the first company to use a Mandarin orange for a soft drink and that might be a clue. I worked at a depot at Cannock in the West Midlands but they closed that so I was asked by Chris Rainsford who was then the Managing Director of Mandora if I'd move to Mansfield. Geographically at least, this was a big thing; although it was only 80 miles away the grandparents thought they might never see the children again! The first time my parents came to visit us they prepared a flask, they were checking tyres, it was a big trip – now, of course, to me, it's just an hour, but back then it wasn't. 1979 doesn't seem a long time ago but attitudes then about travel were very different. It was a long journey – as well as the flask they had to stop at a Little Chef en route and have a meal. Working at Mansfield seemed strange at first but as a salesman I found it easy to be successful. The reason was that everyone seemed to know the brand, Mandora, and with its association with the Brewery you couldn't go wrong; the Brewery's reputation was so good. It was totally different from what I'd done before. I was trying in the Birmingham area to sell a product that not many people had heard of. But when I moved here everyone had heard of it. Although the West Midlands was only 80 miles away, back there they didn't have regional beers. The dominant brewers then were Bass, Mitchells, Allied Breweries and Ansells. Wolverhampton and Dudley Breweries and Banks were very small concerns. They were based in and around the Wolverhampton and Dudley area, and although they were just 9 miles from Birmingham, no-one had heard of them, unlike the situation from Worksop to Mansfield. Strange how the

wheel turns; 23 years later Wolverhampton and Dudley buy Mansfield Brewery. Mansfield Brewery bought a company in Evesham called T. W. Beech. I worked in Evesham for a couple of years.
"When the Brewery built the new brewhouse, John Hings, the Managing Director at the time, needed to fill it. At the time the Brewery owned 350 pubs, but they'd built the brewhouse for the future, on the basis that if you're going to build one, make it a big one. But with 350 pubs they didn't have enough volume to fill it. So, the way that we did that was to bring in contracts from Tesco and Safeway, Asda and Sainsbury's and the take-home volumes went from 3,000 barrels per annum within 2 years to over 200,000 barrels per annum. Throwing these big orders at the brewhouse – 70,000 barrels, 50,000 barrels – it caused quite a bit of mayhem. But the idea was to hold on to this business and still do all the other things we wanted to do. This led to the most exciting time of my life, when we began exporting overseas."

TO RUSSIA, WITH LOVE.

"I arrived in Moscow in December 1993. The Berlin Wall had already come down. I was told before the trip that I needed stout boots with thick soles, and especially a good, heavy top coat. So, I went to Dennis Foster to see if he'd pay for these - after all, I was going to Russia for the company. Dennis said no, but then eventually he spoke to Richard Chadburn and he agreed that the Brewery would buy the gear. Of course, I had to get the receipts. I went to the Baltic States first. It was minus 8 degrees there and I thought 'This is cold..' and then I went to St. Petersburg and it was minus 15. But when I got to Moscow, it was over minus 20 degrees and if you added the wind chill factor it was probably minus 40. Your lips go numb; you can't speak whether you want to or not because you're so cold. We did some fantastic business in Russia. It was quite exciting. We sold beer to America as well and we were learning a lot about exporting. For three years the Russians were paying us up front and business was good. Then the board decided that we should treat these people as proper customers and give them credit facilities. Three months later, the rouble crashed. We ended up with quite a lot of stock in Moscow and quite a lot of beer on its way to Russia by sea. It was a very painful experience but then, everyone who traded in Russia had the same painful experience. I went out to Russia with the two Russians, Simon and Anya, who were our sales agents, a husband and wife team. I learned a very important lesson which I still live by today. I was overwhelmed by their generosity and hospitality. Whilst I was in Russia, I was, to them, the most important person there. Prior to that, looking back, if someone came to visit me in Mansfield I'd find an hour for them. No matter where they came from. My boss was the same; we'd see people as an interruption. We must have given that impression. But when I saw the opposite of that, I was quite ashamed of myself. From that day to this I have never behaved like that again. Since then, if someone takes the time to visit, they would never know from me that they weren't the most important person there. The Russians were amazing; very much like us, but so hospitable. I was told before I travelled by the man at the Foreign Office not to dress too smart and to look out for 'the honey trap'. This, apparently, involved very good looking Russian women who might see this well-dressed Westerner and sidle up to him maybe with something to tell him. I thought straight away – 'I'm interested!' James Bond look-alike? I took a big file marked 'Top Secret – Do Not Open' with a CIA label – not a bloody sausage! I thought I might complain to the Foreign Office. Dennis Foster never wanted to travel, but I got to know my two Russians very well and it came to the point where we wanted to draw up a contract and they wanted the exclusivity of a certain geographic area. To be perfectly honest it meant not a jot to Mansfield Brewery. They could have said 'the rest of the world' and it wouldn't have mattered. So we had this meeting with the Russians at the Brewery and Dennis Foster gets this huge map and lays it out on the table. It became like one of those wartime territorial conferences. The Russians were pointing out the areas on the map and Dennis is saying, 'Oh, no, you can't have *that* bit..*we're* having that bit'… trouble was, Dennis didn't even know what 'that bit' was! But Dennis just had to negotiate – he was that sort - he would have negotiated with the milkman."

"The crash of the Rouble was the end of the Russian business. But although they had a population of 256 million – similar to the USA, they didn't drink a lot of beer. They drank vodka. So all the major brewers saw it as having huge potential. But it was fantastic business while it lasted – and it taught us all a lot of lessons. As for the American business, that's running to this day. In fact Colin Stump, Commercial Director of Mansfield Brewery, asked me if I would go and live in the states and run the business from there. I was contemplating that when the bid for Mansfield Brewery was being made. I contemplated going to America but decided not to. In America, you had to take the word 'bitter' away. We had to call it Mansfield Ale. If you used the word 'bitter' that's what they thought it might taste like. I learned early on that if you introduced yourself to the Americans as 'Mansfield Brewery' it meant nothing because they had no idea where Mansfield was. But, for instance, if we'd called ourselves the 'Sherwood Forest Brewing Company' they would have known straight away, geographically, where we were. We did actually think about doing that. I liked the idea. But I was often told off for having a 'British point of view' when it came to sales. I used to say to the guy who we'd employed as an import agent 'How are we doing in this state or that state?' and he said to me 'California is three times bigger than Italy and it's the 5th largest economy in the world – and you want me to do every state? Do you know every customer in Britain?' Then I realised that we couldn't possibly get to know every potential customer in America. It's difficult to grasp the scale of America. It's so huge. But I used to come back from America quite refreshed. They traded differently to us. We had a scheme, 'serving quality' and I'd been selling the beer with this in mind. Our American agent said 'What's with this 'serving quality'? Why do you need to have this?' I explained that it was to get people to behave differently and to respond to customers, etc., and he just said 'Don't employ them in the first place! Why would you want to employ somebody who needs to be taught 'serving quality'?' He thought it was a huge negative when I thought I was selling a positive. He said 'You go to any bar or café in America and you'll be served by a nice girl who works hard, is pleasant, and gets a tip. She doesn't need no 'serving quality' course to teach her how to talk to customers!'. I also got told off for telling this guy what he might earn on a percentage margin. He had a point."

PINK ELEPHANTS & PURPLE GIRAFFES

Terry: "I got the opportunity at age 55 to take an early retirement pension, which I did. I then worked part time 2 to 3 days a week for the new owners, Wolverhampton & Dudley. Some people say to me 'Terry – how can you work for *that* company?' I reply that Wolverhampton & Dudley are OK – I've no problem with them at all – they're good people." Bob continues:

"The only difference is, they're *not Mansfield Brewery* – so in many ways I think, as workers, we've been spoiled. It's just a different culture, yet Wolverhampton and Dudley, exactly like us, know all their customers by their first names, they all get along with each other, just the same. I think they're a good company, but the experience we had at Mansfield was different."

"When you think about it, the takeover, Wolverhampton & Dudley just had to go for it. They couldn't miss it. But I remember when I moved up here in 1979, all the welfare clubs – Clipstone, Blidworth, Forest Town, there was only one beer on the bar – there might have been upwards of 20 or even 30 pumps on the bar – all of them Mansfield. They did consider Home Brewery (Nottingham) but even that was thought to be almost 'foreign'. You could draw a ring around the area – you had all the miners' welfare clubs stocking Mansfield, the Nottinghamshire Miners' Pension Fund investing in Mansfield Brewery – it was cradle to the grave almost. In fact, I might be exaggerating, but it was almost that kind of pride Irish people have in their Guinness.

This was the local brew, the local employer, the local company. You

didn't need to have any advertising campaigns around Nottinghamshire telling people to drink Mansfield beers. It was their drink. The miners' welfare clubs were a fantastic organisation. I remember when we were about to go into a TV advertising campaign to sell Mansfield Bitter further afield. The advertising agency sent this bloke up from London; typical media type of the time – he had a pony tail. He got nick-named the 'Genie in the bottle'. So he went out on a research trip around the welfares talking to miners and one of the questions he'd ask was 'How much do you drink?' And the answer would come back, 'Oh, 14 pints.' So he says 'Wow. How do you spread that across the week?' And the lad from the pit says 'Across the *week*? No, lad, that's at dinnertime or after a shift!'
The ponytail couldn't comprehend it. To southerners it was an alien concept."

"There was very little the Brewery wouldn't support. Brass bands, welfare clubs, you would go off to Hucknall Miners' Welfare on a Saturday night and there'd be something like the biggest marrow competition and the Brewery would be involved. It was a social, community commitment. Pensioner's parties, children's party. The Brewery was a benefactor to everything. They were a community player. Robin Chadburn once said to me 'We're not only in this business for money'. I said that I was, but I knew what he meant. If you were running a football club on a Sunday morning and you were short of strip or equipment, you could write to the Brewery and they'd look kindly on you."

Terry: "One of the Brewery's customers would even hold meetings in the Company boardroom – they'd let people do that. But the chap was a good customer. When I started in 1977 Richard Umbers was Sales Director. Years later when Richard was leaving Dennis Foster came on board. He eventually took Richard's job. He was a lovely bloke whose bark was worse than his bite but he could be a bit of a bugger at times. Stories about Dennis are legion, but he was a good man. If you knew Dennis he was alright. I remember John Bennett, a salesman who used to look after our miners' welfares, he took Dennis to a committee meeting at Blidworth Miners' Welfare. Dennis was a big stocky chap, around 20 stone with a forceful character to go with it. He goes pounding, marching into this committee room and thumps the table, sat down on a chair and the chair collapsed and he's down there with his chin on the table. But he didn't get up – he just went on and conducted the meeting from down there!"

Bob: "The directors used to get around as well. They'd go to clubs and pubs, play darts, take part sometimes in meetings. This was the way we sold beer. They knew all the landlords and the stewards at the clubs. And there were quite a few characters among them."

Terry: "Yes, and they'd get up to all sorts. I remember one club – better not name it though – and I'd been worried because they weren't shifting our Marksman lager. I said to the steward – 'Ron – what are we going to do about the lager? We're really struggling here.' He said 'Don't worry – we'll sell it. No problem – come and see me next week and see how much we've sold.' So, I went away, and a week later I checked with our order office to see how this club had done with the Marksman. They said 'We can't believe it Terry – they've had four 22s this week when they normally have one a month.' So, I went to see this Ron and said 'Come on, Ron – how are you doing this?' He pointed at the Carling Black label pumps and said 'You see these – Carling – well – it's Marksman. They wouldn't drink Marksman, but they'll drink this one!' So, he'd simply switched pumps. That's an example of a steward doing the salesman a favour!

Bob: "The joke at the time in the welfares was that the President and Secretary of the club would stand up at the committee meetings and say 'We're in a terrible financial plight, but we're very lucky because the steward says he's going to help us out. But although that was a joke, there really were stewards who could do it."

Terry: "Yes, that's true – I remember one steward who actually offered to help a club out by lending them money to pay the debts. Welfare club stewards were a special breed…The man who followed Harry Ridley in the Wines and Spirits Department was Derek Evans. I remember we were at the Licensed Victuallers' Association Ladies Week – an annual event – in Skegness. We were there for a couple of days. We'd been out for a few drinks – as we used to in those days. We were sat in this hotel and I said 'Derek… I've jut seen a pink elephant walk past…' Derek looked me up and down and said 'I'm going to get you to bed…' But then a purple giraffe walked past. Apparently the circus had come to town and they'd painted all the animals different colours and paraded them along the front. It became the joke of the week – pink elephants, boozing sessions…'

Bob: "I remember something similar. I was sitting in the reception area of Sainsbury's head office in London. It was wintertime, and I was sitting talking to other people. Of course, most of the people visiting Sainsbury's head office had come from the south east so they hadn't travelled very far. This bloke turned to me and said 'Where have you come from?' so I told him; 'Mansfield, in Nottinghamshire.' He said 'Oh…that's a long way. I heard about Mansfield on the news yesterday.' He told me that a circus had come to Mansfield and overnight a camel had slipped on the wet ground, his front legs had parted, and he'd died because of the strain on his heart – apparently the camel's heart is up there in the rib cage and if their front legs splay apart, they die. Then the chap says 'I thought it was the shock of being in Mansfield that had killed him. I mean, I've never been to Barnsley, but I know what it's like…'

Terry: "When Harry Battersby was Free Trade Manager we used to go out a lot. In those days we'd work four nights a week minimum and quite often Saturdays and Sundays; visiting clubs, pubs, attending committee meetings. The working hours weren't a problem for us because the company was so good. Harry was a musician in brass bands and after the war he'd actually played in a successful band called The Debonaires. Harry was noted for his music. Wherever you took him, especially to a club, they used to look forward to him getting up and giving a turn, singing a song. He was absolutely fantastic. I remember calling for him one night. You always had to bang loudly on the door because he'd be practising away on his euphonium. He'd leave the curtains open so he could see you arrive. This night we had

to visit a pub and I suggested that he bring Brenda, his wife along. Sadly, she's no longer with us but Brenda had a bit of a reputation for enjoying a drink. So we went to this pub.... I don't know how many she'd had but we're coming back in the car and Brenda's laid back in the rear seat snoring her head off. We pulled up at Harry's house and Harry got out. He says 'See you tomorrow, Terry...' and I looked to the back seat and shouted 'Harry! You've forgotten something!' He says 'Looks like you're stuck with her – I'm not having her in here!' So I had to get out and shake her – 'Brenda! Brenda! Wake up!'. There was no way she was going back with me!
Harry was a lovely bloke. He was killed in an accident on the motorway. We'd been attending a meeting on a Friday afternoon. Harry, being the sociable type, would usually say before we all parted 'Right, meeting's over – lets have a half in the bar before we go'. I recall it was at the Dukeries in Edwinstowe. Yet this time he was in a hurry – he had another meeting to go to up in Blackpool so he wanted to get off. There had been a multiple pile-up on the motorway and he saw these two ladies who'd become stuck in a car which was part of this concertina of an accident. He parked his car up and went over to help them to get out and make sure they were safe and a car hit him and killed him. He was the only person killed in that accident yet the only person not involved in it. But that was Harry – a proper gentleman, always helping people."

Bob: "After I moved up here I always remember that one week I was having some time off to drive down to the West Country. Harry knew I had a smaller car than his and he said 'How many are going?' I said four. He said 'That'll be a tight squeeze in your car – take mine for the week.' He had a larger car than mine and he swapped with me. I thought it was quite a gesture because I hardly knew him at that point. He was a good man. There were plenty of good men like him. John Bennett, Jess Musgrove, Michael Brown, for instance."

Terry: "Yes – Michael Brown appeared in a film - *The Charge of*

The Light Brigade. That's because he'd been in the Household Cavalry – he was good with horses – so he actually got a chance to ride in the charge. Another item worth a mention is Fred's Kellar. Regularly, those of us who were in the Brewery on a Friday afternoon used to meet up in Fred's Kellar. Basically, it was the Brewers' sampling room. It was a nice, cosy little bar run by Fred Hensleigh. But you only went there by invite. You sampled the beer that had been brewed that week, and even if there were any beers which had been returned or were doubtful, you sampled those as well. Sessions in Fred's Kellar were quite an event."

1992: Bob O'Meara with fellow long service award recipients (l-r) Syd Monks, (Richard Chadburn presenting) John Lewis and Mike Cleator.

END OF AN ERA

Bob: "If they had the chance most of the people who worked at the Brewery would start back there tomorrow. I often tell people that some of the big people I dealt with – people like Tesco, who were very hard to deal with – they were very tough negotiators, when I told them that there was a bid in for Mansfield Brewery, they said 'Look – we don't want that to happen. What can we do? Can we send in our accountants – is there anything we can say to Price Waterhouse that will help?' And Tesco weren't the only ones – many other big accounts offered help, anything that might have ensured Mansfield remaining independent. When the court case came and we all went down for the hearing, they were quite surprised to see such a lot of people turning up at the court hearing. We had a bit of a deliberation with the Judge but we were told it was quite OK for us to be there. Mentioning no names …there were three barristers and there was a guy next to me from Cellar Services at the Brewery who listened to

this barrister put his case in the morning. Anyway, we went across the road for some lunch and one of the blokes from Cellar Services said to me 'I don't think we've much chance of winning this case – in fact listening to that bloke we haven't got a good case at all..'
I looked at him and said 'That was *their* Barrister, you silly prat!" He looked at me and said 'Oh... I though he was *ours*!' So, he'd spent all morning thinking the Barrister was pleading our case."

Terry: "Yes, working for Mansfield Brewery was an unforgettable experience. I doubt there'll ever be another company quite like it. I enjoyed every minute of my time there'.

❖ ❖ ❖ ❖

1997: Jaclyn Bateman and Bob O'Meara promoting Bateman's and Mansfield beers in Poland.

1992: Managing Director Bill McCosh (left) presenting Terry Johnson with his long service award for 15 years.

Chapter Ten
THE LAST BARREL

The Fight to Save Mansfield Brewery

'Some of those who attended the hearing were appalled at the full majesty of the British legal system in full flight as it waded in against the employees' case with some scurrilous devices, an example of which was the unchallenged claim that a great number of Mansfield Brewery's public houses still had outside toilets'.

The historical, financial and legal basis of Mansfield Brewery was an unusual one. Despite the fact that the company was highly successful, turning out increased profits over the years, and always remained in the vanguard of new developments in the industry, those with a more avaricious view towards profits saw the way the company's historical background had developed as a problem.

Innocent and both mentally and legally incapable of making a Will, Frank Baily Thomas, for whom the Baily Thomas Trust had been created, would own half the company until his death in 1997. With no living relative, the potential outcome of his death could be, therefore, that whatever fortune he had, which was considerable, would go straight to the Crown. For the Brewery, this would have a devastating effect on the company's financial structure. Frank's father, Captain Munro Thomas, had passed away in 1970, and there appeared to be no known relatives to which Frank's legacy could be passed. However, the intrepid Jock Nangle would eventually trace a long-lost brother of the Captain's. But the aged Hector MunroThomas, estranged from his brother for over five decades, had no wish to become involved.

But the aged Hector Munro-Thomas, estranged from his brother for over five decades, had no wish to become involved.

However, a fortuitous change in the law diverted Frank's fortune away from the Crown's coffers. This permitted, through the Receiver, the Court of Protection to approve the making of a Will by a person previously considered incapable of making a Will, provided his affairs were under the Court's jurisdiction. So it was that in 1970 in the Chancery Division, a Judge gave his approval for a scheme which allowed for most of the unfortunate Frank's fortune to be divided into a fund for the benefit of the Employees – the Baily Thomas Provident Fund – and a charitable fund for research into mental illness. Thus we can look back to 1970 as a time when a fair and balanced Court, based on the information provided by Jock Nangle, displayed a laudable understanding of the fact that what hung on their decision was the continuation of a very successful local company which formed a very important part of Nottinghamshire's social infrastructure. Three decades later, in a far less compassionate corporate climate, such benevolent notions could easily be dispensed with, and the name Nangle would feature again in a slightly different light.

From the staff's vantage point, the memory of the bitter period of struggle and disappointment leading up to MBC's demise as an independent company in 1999 demonstrates a period of confusion, claim and counter-claim. The dedicated workforce, from management down to the dray men, would become desperate to fight to keep the company they loved and the jobs they enjoyed so much. They were suddenly faced with the prospect of examining the complex legal structures which lay ahead. Interpreting these was in itself a small minefield, leading to many internal disagreements and misunderstandings whilst in the financial corridors of power, in London and Wolverhampton, capitalism's financial and legal infantry prepared for the battle.

Senior Manager Paul Cooper, who spent much of his time at Mansfield Brewery in charge of the company's sophisticated IT equipment, recalls the day the announcement was made about the Brewery's potential take-over.

"It was mid-summer in 1999 – Mike Cleator's birthday as I recall. John Else, the Brewery's Head of Personnel, called for the senior managers to meet at 5 p.m. At the meeting, John outlined his belief to the gathering that the Trustees would not be able to sell their shares if this take-over bid advanced any further because of the conditions in the Trust Deed. John's emotional presentation explained that the Trust Deed had stipulated that Mansfield Brewery would be kept independent. There was some disagreement about this and it became the source of some argument. For the Trustees, their chance to sell was based on their interpretation of a clause in the Deed that the Trust should retain the shares unless circumstances made it desirable to sell them.

But things moved on, and we needed to see a copy of the Deed to clear things up. We actually had to 'lean' on the Public Trustee to even give us a copy. Luckily John Else had a legal friend, a solicitor, up in Leeds, and we went up there to see her. It was a very disappointing experience, because this solicitor concerned was acting for some other company, which was operating in similar circumstances. Her advice to them was that they shouldn't have a single company trust, that they should sell their shares and acquire a multi-share portfolio."

John Else, feeling very dispirited, went away on holiday. In his absence the opposition to the take-over continued to build, with Paul Cooper and other managers. After approaches had been made by Terry Johnson, they visited an important body in the Brewery's commercial domain, CISWO, the Coal Industry Social Welfare Organisation. They used to be responsible for all kinds of social support, such as the Miners' Welfare Clubs, etc., and therefore obviously over the years the Brewery had a close relationship with the organisation's beneficiaries. There they were introduced to a helpful lawyer, Joseph White. He was on a retainer from CISWO and thus was able to give the Brewery staff advice as part of his everyday work for the organisation. Mr. White advised that there were avenues open to the staff through which they could tackle the problem. Paul takes up the story.

"Mr. White steered us into how we could legally pursue the matter. CISWO had no desire to see us go down. We were looking for a way to support the emotion at our meetings which was asking the question about the Trust – if the Trust was to be for the staff's benefit then how could they sell the very shares which would put us out of business? The first thing we did, rather than just rely on what we knew from CISWO was seek the opinion of Counsel on this. Mr. White went down with us to see a Q.C., Mr. Picarda, at Lincoln's Inn. That cost us about £5,000. At this point I think we raised the money from contributions from senior managers.

Another potential plan was that the board might organise a management buy-out and, with loans, fund the purchase of the shares.

At one point we had a meeting with the Public Trustees to try and elicit from them what the basis of their action in selling the shares was. I even gave them a financial presentation. But at this meeting we discovered that the Trustees had a report from the big accountancy firm, Ernst & Young".

The general thrust of this report, which covered lots of economic points and financial views, made a case which some might argue culminated in the somewhat surprising suggestion that there was no future in the brewing industry for a company like Mansfield Brewery. In a previous incarnation, as Ernst & Whinney, this was the very same firm which had employed Jock Nangle – who had remained a Trustee of the Baily Thomas Provident Fund. Three decades later, it would be another employee of Ernst & Young, Jock's son, Jocelyn Nangle, who would play a part in Mansfield Brewery's eventual demise.

"This report had provided the Trustees with the basis for their decision to sell shares. The general feeling we got was that we were not so important – we were just 'employees'. Humphrey Mather, the voice of the Trustees, said 'I hear all you're saying – but we have this report… the only way we might change our minds is if you present us with a case based on what you're telling us.' So I suggested therefore that we had a report of our own."

"Joseph White had already outlined to us that he thought we had a case. When we got the opinion of Mr. Picarda Q.C. , which outlined his view that the Trust was not acting in the beneficiaries' interests, we sent that to the Trust. I remember that at this point even bigger lawyers became involved, because after our meeting in London with Picarda, Joseph White, John Else and I went into a Wetherspoon's pub which had previously been a bank. Joseph advised us to seek the support of one of London's larger legal firms. John and I decided to stay overnight to see if we could get further with our case. We telephoned a big legal firm, Linklaters."

The following morning, armed with Mr. Picarda's opinion, John and Paul met with a top lawyer at Linklaters. They were advised that three letters had to be written to three important people in the case. These letters would outline Picarda's arguments. Each letter would cost £1,000. However, they were never charged for these, as Paul explains:

"Within a couple of days we got a phone call from Linklaters saying that they couldn't act for us because they had a 'conflict of interest within the firm' . They did the letters, as arranged – and didn't charge us – it all seems very odd in retrospect as we never had a full explanation as to why Linklater's wouldn't act for us. Had they been 'got at'? They did, however, recommend us to another corporate lawyer, Herbert Smith. By this time the fight was into its third month."

John Else recalls that they had become acutely aware that all this legal wrangling was to be costly, and a fund was organised. The fighting senior managers at MBC were committed to paying a minimum of £250 each. From the employees, contributions ranged from 2p to £500. Neil Tomlinson and Richard Meadows had approached friends and associates in the brewing industry. Eventually the fund gathered over £70,000. However, much of this was to be spent on the management's own report, which would counter the Trustees' version of the situation, prepared by Ernst & Young. The management team used a financial consultancy, L.E.K. This weighed in at a hefty £60,000. The fighting committee's sole hope was that it would persuade the Public Trustee that selling the shares and allowing the take-over was not the thing to do. There was undoubtedly pressure at that time from the Charity Commission to charitable trusts that they should not have single company holdings and as a result they should diversify. I could therefore understand that anyone on the charitable trust who was receiving such advice would feel that they needed to diversify their Mansfield Brewery holdings. (However, it has to be noted that the Charity Commission supported the setting up of the Baily Thomas single company funds in 1971, which was long after they first recommended mixed share portfolios).

The management group, through Mansfield's Labour M.P. Alan Meale, sought a meeting with the Charity Commission. But there was a downside to this. The meeting took so long to organise that in the waiting period the Charity Commission had already made its mind up over the Mansfield Brewery issue. Any attempt to influence their deliberations would seem futile. The meeting with the Charity Commission ended in failure.

The battle to save the Brewery would eventually be won or lost in a court case. Public attention to the fight was widespread, and a march was organised through Mansfield centre. A well-attended public meeting was held in the town's Civic Centre, with speeches by MP Alan Meale, the Bishop of Sherwood and various councillors. Both BBC Radio and Central TV were there, and a mail-out to the shareholders was organised, asking them to stay with the campaign by not accepting Wolverhampton & Dudley's offer until after the court case was decided. What was needed was more than 25% of the shareholders to get behind the campaign to save the Brewery.

"We almost made the required percentage," says Paul, "but ultimately it didn't work. We had been through all the grounds upon which we outlined why we thought the trustees were making the wrong decisions, but what we didn't actually do in presenting the case to the court was say 'This, from our perspective, is what we would like you to decide.' By the time it reached the court the whole take-over process was well down the road. We had gone along with our solicitor's advice, but we were well out of our depth, really."

Some of those who attended the hearing were appalled at the full majesty of the British legal system in full flight as it waded in against the employees' case with some scurrilous devices, an example of which was the unchallenged claim that a great number of Mansfield Brewery's public houses still had outside toilets. The overall impression given by capital's be-wigged champions was that MBC was a creaking, unsophisticated, provincial organisation, almost in the cloth cap and whippet tradition, long overdue for modernisation. Nothing, sadly, was further from the truth – this had been one of the Midlands' most successful and sophisticated independent businesses with some of the finest properties in the industry. Yet such crude legal broadsides had the desired effect.

That December, 1999, was the end of the line. A century and a half of brewing at Littleworth had come to an end. It was a terrible Christmas present for an already demoralised staff.

Paul Cooper was one of the last men to switch off the lights in the building the following year. Head of Personnel John Else's lights went out in February. Paul's pride and joy, the intelligent and state-of-the-art computer systems he had helped to develop to run the Brewery, were suddenly faced by W&D's slightly less technical approach;

"This grieved me the most. As head of IT for a large part of my employment at Mansfield, I found all the systems which had been installed to successfully run the business just being turned off or phased out simply because they were more advanced than anything W&D had. The warehouse system just got turned off and they replaced it with blackboards! They weren't on our wavelength."

Six years after the premises at Littleworth ceased brewing, apart from the main office buildings, the huge site remains, a crumbling, derelict, prime, brownfield land waiting for a developer. The warehouses and numerous structures used in the brewing process still stand, albeit in a sorry state. The once proud brewhouse is falling to pieces. Mansfield Brewery was at the heart of the community. No doubt, whatever becomes of the site – a hotel, a leisure facility, housing, those few golden acres will always be populated by the ghosts and, needless to say, spirits, of generations of very happy employees whose product, over a century and a half, brought joy to millions of working people throughout the Midlands and Yorkshire. Of course, the brand, Mansfield Beer, remains. The pubs still bear the legend, 'Mansfield'. But the original well from which the sweet nectar flowed has dried up. In Hull, the same situation arose at Silvester Street, the site of the once-thriving Hull Brewery. Yet there has been good news for Hull's beer drinkers. Some ex-Hull Brewery employees have got together to form a company of the same name, using original Hull Brewery recipes. So those of us of a certain age who remember Hull's fine old mild and bitter can indulge in some genuine nostalgia when visiting the city.

Sadly, the pub trade, which once attracted so many colourful characters, has also suffered from the rapacious advance of 21st century commerce. One ex-Mansfield Brewery manager, who will remain anonymous, illustrates the way the pub trade has presented a new challenge;

"To give you a flavour of what's happening in the industry now, I looked at a pub recently which had come up as a lease-rental. I thought I'd take a look at it and see if it was worth investing in. But I worked it out that they were wanting £52,000 per year for lease rental of the pub. And that's not including a figure for fixtures and fittings. It's an ordinary pub, nothing special. Now to cover that, if you ran this pub, you would need to be taking a minimum of £14,000 per week out of a lease rental tenancy. The most this pub has ever done in its financial heyday was £7,000 per week – so how on earth is a landlord going to cover £14k per week - £2,000 per day? And this is why they can't get decent people to run pubs. The situation is virtually impossible. The days when the pits were open and a pub would do five barrels on a Sunday lunch are long gone."

However, despite the crass, infantile fashion, currently beloved by youth's baseball-capped legions, of drinking bland, over-priced inferior beers with American branding and expensive, technicolour alco-pops direct from the neck of the bottle, good cask-conditioned beer is still available in a great number of pubs. Mansfield Brewery Plc may be a memory, but it's a good one. We can be proud of what went on down at Littleworth. So remember – the next time you have a pint of Mansfield beer – don't cry into it.

❖ ❖ ❖ ❖

All things must pass: even these proud coppers on the copper floor of the brewhouse, pictured by Vicky Phillips in 2000.

One last, lofty view: the Brewery Yard, 2000, captured by Vicky Phillips.

A PINT OF MIXED:
A Random Miscellany of Memories:
Bottled Recollections from A Brewery's Past

Down the exciting and productive decades of the 20th century all kinds of snippets of news and stories have stood out from the backdrop of the everyday business of brewing in Mansfield. From the late 1970s to the very end of life as it was known at Littleworth, the staff and workforce were kept abreast of developments both large and small with a series of company newsletters which were issued under such mastheads as *The Marksman* and *The Brewery Independent.* What these publications reveal is a company with a genuine family feel and a sense of benevolence towards its staff. The further back the newsletters go, the more that feeling of 'belonging' is prevalent. Engagements, marriages, births and deaths along with sporting triumphs all share column inches with the higher-octane stories emanating from the boardroom.

Not all the stories included in this random selection come from *Marksman* or *The Brewery Independent.* Others have surfaced by word of mouth or have been selected from various media who have shown an interest in the Brewery along the way. But taken as a body of work, for future historians, *Marksman* and *The Brewery Independent* will provide a unique insight into how business and employment worked in the last decades of a century of change. One could argue that, in the 21st century, it would be hard to find a comparable corporate atmosphere as amiable and conducive to good worker relations as that which is evidenced by the pages of the newsletters. These employees worked well because they were treated well. Unlike today's grabbed sandwich and bottled water working lunch environment, the workers at Littleworth had time to get to know one another and their employers. Service and loyalty were respected.

If you once worked for Mansfield Brewery, there may be the odd story in our miscellany here which will trigger a happy memory. If you weren't a Brewery worker, this selection will, none the less, offer some indication as to why local people were so happy to work for the company, which inspired many to notch up impressively long service records.

These random entries also remind us, somewhat sadly, of how much life has changed for employees in today's aggressive corporate world, where having the time to read a newsletter could well be regarded by the new time and motion 'suits' to be an interference with production. The bright and breezy, mostly optimistic tone, especially in the early 80s stories, also makes one realise the unbelievable impact the Brewery's end must have had on this dedicated and happy workforce.

Peace, Love and Ale;

The 1960s.

In 1963 Claude Chadburn retired, aged 76, to become Mansfield Brewery's first President. The new Chairman would be his nephew, Robin Chadburn. 1963 gave us the worst winter on record since 1881, and on the social front, unemployment was on the rise. This had little discernible effect on the company's profits, however. Idle workers have more time to forget their troubles over a pint. The Brewery pressed ahead with major re-development schemes at the Wheatsheaf in Bakewell, the Red Lion in Farnsfield and Skegby's Forest Tavern. At this time laws were passed to legalise gambling – the dawn of the pub 'one-arm bandits' had arrived.

New Blood – and Tragedy – on the Board

Robin Chadburn had discussed the possibility with his uncle and Jock Nangle of appointing the tenant of Newstead Grange Farm on the Papplewick Estate, the efficient and ebullient David Weston, as Sales Director to the Brewery's Board. Weston was married to Claude's youngest daughter, Anne, and was an efficient farmer with great charisma. Weston was pleased to accept the post, and was soon attending the Board's fortnightly meetings.

Unfortunately, whatever potential David had to offer – and it seems boundless – the Brewery were not to benefit, as he died after being thrown from a horse. This left Robin Chadburn with the quest to replace David Weston, and his choice seems to have been a wise one. Tony Scruby was an ex-military man who had served as a Brigade Major with the 8th Battalion Sherwood Foresters. He had seen action in both Burma and the Mau Mau conflict in East Africa – and to boot he was a qualified solicitor specialising in commercial law.

Late in 1963, following discussions with Jock Nangle and Robin Chadburn, Tony Scruby joined Mansfield Brewery's Board.

AN ARCHITECT'S CHALLENGE

Here, in licensed premises, the architect always faces his biggest challenge, because in pubs the bulk 75% of the space is taken up with accommodation for the landlord's family, with only 25% left for drinkers. In the club world, the ratio seemed to end up as 60% licensee, 40% drinkers. That said, the profits rolled on.

DRAUGHT BEATS BOTTLES – and something for the Ladies…

In late 1963 to early 1964 party-sized cans of Mansfield Beer grew rapidly in popularity. The party can gave customers the enjoyment of drinking draught type beer at home, and provided a drinking experience similar to draught ale from behind the bar. More and more women were demanding their right to go out for a drink. It was a major social turning point, as for the first half of the century the pub had been seen as the domain of the man of the house. He was happy to prop up the bar in a spit-and-sawdust atmosphere, but the new breed of women demanded a little more comfort and sophistication. The Brewery as ever responded to consumer demand and new innovations such as comfortable furnishings, carpeting and cloakrooms became the norm.

NEW PUBS AND NEW DEVELOPMENTS

On 4th April 1964 President Claude Chadburn proudly opened the new Rushley Inn in Mansfield, and later that year repeated the ceremony for the new White Swan at Mansfield Woodhouse. As a reminder of the area's mining heritage, at Codnor the Glasshouse Inn began to suffer from subsidence which would require this popular pub to be rebuilt. Elsewhere the modernisation continued apace as the Devonshire Arms, Sutton, Staveley's Albert Inn, Shirebrook's Great Northern Hotel and Gate Inn, along with the Wellington Inn at New

Whittington all received a thoroughly modern make-over.

MANSFIELD RESISTS CHANGE

The Mansfield brew continued to be as popular as ever, but there was always the lurking danger of a take-over by the big brewing groups. When this happened in other areas, drinkers soon had plenty to complain about as their well-established local brews vanished overnight, to be replaced by whatever national concoction the marketing boys of the brewing giants were plugging at the time. Had the corporate leviathans got their way, then that much vaunted buzz-word so beloved in the coming Thatcher years - 'choice' - would have had a hollow ring because we could easily have been limited to a world of Watney's Red Barrel and mass-produced insipid lager.

But Mansfield beer was the beer the people wanted, because there was a consistency in quality. The company strode on, confident and independent, with a beer production rate which was running at twice the national average. This continuing surge in production saw the men at Littleworth taking new decisions – equipping No. 4 fermenting room and completing the conversion of fermenting room number 2. Even the Chancellor's predictable inroads into the drinker's life in April 1964 – a 10% increase on Excise Duty on spirits, tobacco, wines and beers with a 15% surcharge on wine – couldn't stop the Mansfield juggernaut as profits by November were £661,152.

THE TANKS ROLL OUT…

To most beer lovers in the mid 60s the very idea of ale being delivered not in barrels but by tanker was an unforgivable heresy. Yet this new method of delivering ale meant less work for draymen, and, more importantly, at the clubs and pubs where cellar tanks were installed, the tedium of running up and down the steps to change barrels, especially over a busy weekend, was removed. In any case, drinkers soon discovered that the beer delivered by this new method was in every way as good as – if not better than – the old version in barrels.

SPORT TAKES A DIVE

Society had begun to change rapidly since the 1950s and among the nation's youth there were other diversions – fashion, music, etc., which were displacing the previous decade's healthy interest in sport. There was a marked decline in interest in the subject among the brewery's younger staff, a fact which inevitably led to the eventual closure of the Mansfield Brewery Sports Club and the selling-off of the firm's playing field.

BEER AND POLITICIANS

In October 1964 the election was a tight-run thing with Tory Prime Minister Alec Douglas Home losing out to the new Labour PM, Harold Wilson, who scraped into power with a majority of just four. At the age of 91, wartime leader and beloved statesman Sir Winston Churchill passed away. Mansfield Brewery commemorated the great man's life by naming their new pub at Annesley Woodhouse after him – the Sir Winston. Another new hostelry was the Oval Inn at Sutton-in-Ashfield, and the brewery bought the Vernon Arms in Nottingham and the Lord

Nelson at Ambergate. The effect of politicians on brewing continued; the new government decided to introduce a measure which didn't go down well in the boardroom – the Capital Gains Tax. By the end of the year Edward Heath became Leader of the Conservative Party.

THE ONE THAT GOT AWAY

If the police caught you with an illicit 22 gallon barrel of Mansfield's finest cask bitter then the line 'it fell off the back of a lorry' might not go down too well. But in this case, it was the truth. As the Brewery's draymen were delivering beer to the Nottingham Boat Club a cask of bitter made a break for it and tumbled away into the River Trent. Fortunately, on hand were the Mansfield 735 Branch of the British Sub Aqua Club who managed to retrieve the rogue cask from 12 feet of muddy water. The cask itself was worth £70 without its contents, but although it had sustained a dent the beer was passed as OK by the Brewery quality control department. As the Sub Aqua Club held only amateur status, they were unable to accept a cash reward so the contents of a further 22-gallon cask were sold in the Club house on Mansfield's Westhill Drive, and the proceeds went towards an effort to raise £200 for a ship to shore radio.

1970: Jim Wain receives his long service award for 32 years service from Claude and Ethel Chadburn.

Chicken In The Basket & Kipper Ties:

THE 1970s.

Ushered in by the dreamy youth culture of the 1960s, the 1970s seem, in retrospect, to have been that time when a new generation had finally thrown off any lingering, grey uncertainty, a legacy of post-war Britain which had hung around so long. The majority of the 'Flower Children' had grown up, got married, and were eager to enjoy some quality leisure time - and Mansfield Brewery were keen to supply it. It was a time of flared trousers, wide lapels, kipper ties and disco dancing. A new age had dawned. People were going out to the pub more, and a combination of pub food, good beer and entertainment resulted in a boom time for the larger pubs and social clubs.

The Mansfield Brew had become more popular than ever, and the bold expansion of the company as it bought up pubs, built new ones and modernised others demonstrated to any competitors that the Board of Mansfield Brewery were in the vanguard of the leisure age.

THE AGE OF AQUARIUS

This was a period of innovation with licensed premises, marked by two major ventures, one of which was a huge success and the other less so. At Chesterfield, a disused office together with a large disused warehouse was extensively re-developed to become a major cabaret and dining venue. The Aquarius could serve up to 1,000 meals per night, and top flight music and variety acts were booked. All this provided a great night out for Mansfield people and the population around Chesterfield, without having to travel north into Yorkshire or south into Nottingham. A second venue, Jesters at Mexborough attempted to duplicate the success of the Aquarius, but the ratio of the audience to the showbiz performance potential meant that not enough punters could be seated at the venue to justify the fees for more well-known acts. However, the building was very attractively developed and its interior was used for TV productions and frequently hired out for functions.

BOYS AND MEN

In 1970 the company decided to recognise the long service of employees and 26 Brewery veterans were given awards. The list includes men who had begun their employment as mere boys, and it makes interesting reading almost four decades on.

	Years of service
Alf Ansell	38
Ron Allen	38
Freddie Burton	30
Ernest Chamberlain	30
Arthur Chapman	49
Bob Gilliatt	44
Jim Halpin	45
Fred Hensleigh	33
Bill Holland	31
Percy Holmes	47
Doug Kemp	32
Mev Marriott	39
Ray Mellors	39
Les Pearson	37
Henry Peet	42
Ken Perry	32
Jack Revill	32
Percy Revill	44
Albert Rowley	50
Les Taylor	34
Sid Tyler	42
Jim Wain	32
Graham Wilson	34
Les Yates	34

Mansfield Brewery Company Limited
LONG SERVICE AWARDS 1970
Claude W. Chadburn 64 yrs.
Edward Deakin 41 yrs.
Alf Ansell 38 yrs.
Ron Allen 38 yrs.
Freddie Burton 30 yrs.
Ernest Chamberlain 30 yrs.
Arthur Chapman 49 yrs.
Bob Gilliatt 44 yrs.
Jim Halpin 45 yrs.
Fred Hensleigh 33 yrs.
Bill Holland 31 yrs.
Percy Holmes 47 yrs.
Doug Kemp 32 yrs.
Mev Marriott 39 yrs.
Ray Mellors 35 yrs.
Les Pearson 37 yrs.
Henry Peet 42 yrs.
Ken Perry 32 yrs.
Jack Revill 32 yrs.
Percy Revill 44 yrs.
Albert Rowley 50 yrs.
Les Taylor 34 yrs.
Sid Tyler 42 yrs.
Jim Wain 32 yrs.
Graham Wilson 34 yrs.
Les Yates 34 yrs.

YEARS OF ACHIEVEMENT

1972 was a true year of achievement. As well as the Aquarius and Jesters there was the completion of the Bell and Anchor in Scunthorpe's shopping centre. Several pubs were enlarged and modernised, such as Sutton's Crown and Woolpack, the Oak Tree and the Rufford Arms in Mansfield, the Little John at Ravenshead, the Ladybrook and the Ling Forest.

In 1974, the building costs of the Sportsman in Doncaster equalled almost the whole capital of the Brewery in 1901, and building the Fairway Inn at Birley Wood was an even greater outlay. But the Brewery kept boldly moving on and also planned the rebuilding of Mansfield's King's Head.

A CENTURY'S SERVICE

The end of another era was nigh with the closure of the bottling department in 1975 where Percy Holmes and his old school friend Sid Tyler ruled the roost. Percy's father had joined the brewery in 1892 to be followed by Percy in 1922. An experienced horseman, Sid Tyler's father joined the Brewery in 1903. He died in 1925, and his wife began work as an office cleaner, a career which would last 20 years. Young Sid started work at the brewery in 1927, and if we add in the 12 years also worked there by his brother, that brings the combined service of the Tyler family to 100 years.

NEW FACES AT THE TOP

This was a busy time for staff appointments. In December 1979 Richard Chadburn, the son of the Chairman, joined the Brewery as a non-executive director. On May 6th Dennis Foster made the move from John Smith's of Tadcaster to join the company as Free Trade Sales Manager. Richard Marrion – another keen squash player - took up the post as Senior Area Manager for Mansfield Inns, and the

Brewery appointed a new Financial Director, Christopher Rainsford. A new Building Supervisor, Clive Almond, opened a new chapter in his career after ten years in the architect's department of Chesterfield Borough Council. New secretaries Brenda Stanley and Madeline Mellor joined up, and from April 1st Austin O'Donoghue took up his new post as Tied Trade Manager.

Mansfield Brewery chairman Mr. Robin Chadburn, right, receives a gift — one of many to mark his 60th birthday — from brewing director Philip Baker at the brewery offices.

Excitement, Challenge, Reward:

THE 1980s

The dawn of the 1980s was marked by the success of Mandora Soft Drinks. In January 1980 Mandora's Managing Director Tony Morton met Chris Shanks, Vice President of Pepsi-Cola International to secure a new deal for Mansfield Brewery. For a period of two years they were to produce 2-litre bottles of Pepsi-Cola for the whole of the United Kingdom. In addition to this Mandora received no less than three gold medals for their Tonic Water, American Dry Ginger and Cola, plus a silver for Bitter Lemon at the Brewex/Packex National Exhibition in Birmingham.

KEEPING IT IN THE FAMILY

In May 1980, George Powell's wife, Mavis, left her work with a local estate agents and joined the Brewery cellar services department as a clerk. George already had a cousin, Michael Powell, working as a tanker driver at the Brewery. Five years later, the Powell circle was completed at the Brewery when daughter Helen joined the Order Office as a VDU operator.

WILLOW, WEEP FOR ME...

On December 2nd 1980 Philip Baker, MBC's Brewing Director, arrived for a special ceremony at the Willow Tree Inn on Church Street, Pilsley near Chesterfield. Mr. Baker was at this time looking forward to his retirement from MBC the following April after a long career with the company which had started back in 1939, when he joined as a pupil brewer. His replacement would be Michael Parsons.

At the Willow Tree Inn, accompanied by the suitably named Licensees, Mr. and Mrs. Jack and Jean Brewin, Mr. Baker planted a willow sapling as a replacement for the original willow tree which had been planted

in 1875 and had had to be cut down owing to disease.
On the social activity front, the Brewery's prowess in sport – namely, Squash, came to the fore when the 'Magnificent Three', John Hakes, Hugh Wilson and Adam Hyde came third in the National Finals held in Wimbledon on May 4th-5th.
Brewery Chairman Robin Chadburn opened two new licensed premises, The Coopers, on Leeming Lane, Mansfield Woodhouse, managed by Eddie and Mavis Miles, and The Poacher, Radburn Road, Rossington, Yorkshire.

STAR PERFORMER

By the summer of 1981, according to a table drawn up by City stockbrokers Fielding, Newson-Smith & Co., based on the profits of the previous six months period, Mansfield Brewery was the 'star performer', ranking above all other brewers – including the nationals.

A CHRISTMAS BLAZER

Brewery delivery men Albert Mitchell and Roger Walker had a hot Yuletide experience when a piece of plastic sheeting blew into their engine and caught fire as they sped along the M1 with their Christmas deliveries. As the lorry's cab extinguisher seemed ineffective, Albert tackled the sheeting with his bare hands whilst Roger almost entered the realm of sacrilege by knocking off the neck of a polypin (that's a plastic container of 36 pints of bitter) and utilised this as a successful extinguisher. Sad though the loss of 36 pints might be, it was still cheaper than a new lorry…
In fact, this period was notable for fiery experiences. Mandora Soft Drinks also had a fire which began in some pallets on the back of a lorry in the empties yard. Once again, two sharp-thinking employees came to the rescue to move the blazing lorry away from the garage repair area. Chris Sankey moved it first, but as the flames increased, it was realised that the vehicle was parked beneath the overhead

telephone wires. Ron Gregory then leapt into the cab and backed the lorry away into an empty space, but by this time the telephone, telex and burglar alarm wires had all perished, leaving Mandora 'off the air' for 24 hours. The cause of the fire was a mystery....

PROFITS & OPTIMISM

In the early 80s Mandora stories dominated *Marksman,* the house journal of the Mansfield Brewery Company. In the Summer 1980 edition, we see the following snippet:

'As from Friday, 13th June 1980, the company name of Mandora officially changed from R. L. Jones and Company Ltd., to Mandora (UK) Ltd., a name rather more in keeping with the company's activities. The Directors are A. J. Scruby – Chairman; A. R. Morton – Managing; R.A.F. Curtois, J. F. Hings, F. C. Rainsford, FCA.'

In the same edition, the highlights of MBC's Chairman's Annual Statement were outlined thus:

- Sales & Profits up despite soaring costs.
- Profits ploughed back into the business thus ensuring growth and greater job security.
- Inflation reduction only secured by all recognising the need to keep wage increases below current rate of inflation.
- Optimism of the Company's ability to compete successfully against increasingly difficult odds because of the quality, determination and never-ending hard work of everyone connected with the business. Thanks to all concerned.

I KNOW WHAT I LIKE...

The Brewery was also running courses for Licensees and the important aspect of this behind-the-bar education was to ensure that Mansfield's beer and lager 'are presented to the public in the way the Brewery wish it to be and to assist those in charge of the pumps to overcome simple maintenance problems, especially at times when

help might not easily be forthcoming from the Brewery – e.g. on a busy, hot Saturday evening when the pumps suddenly go dry!'
Each course began with an introduction by John Else and a brewing industry film entitled *I Know What I Like'.*

IVY'S SWAN SONG

LIFE AT THE SWAN

One of the nice things about MBC's house magazine, was the regular trips down memory lane looking at the lives of the many characters who had worked for the Brewery, in any capacity, high or low. One such story in the Summer 1981 edition was that of Ivy Rowan, who lived on Keyworth Close on Mansfield's Ladybrook Estate. Ivy had spent 22 years as a waitress at the Swan Hotel in Mansfield. She had started at the Swan at the end of World War II. Her first days were marred when, not quite trained, she attempted to pass through the pub's glass swing doors with a tray of glasses with a smashing lack of success. But she persevered in her career. Her training was undertaken by Captain Ian MacLean, a strict but kind MBC man who tutored her in how to lay tables, to carry correctly and an early form of 'customer relations'. Ivy often served the Brewery's top brass when they dined at The Swan, and particularly remembered the Chairman, Claude Chadburn, who was very kind to the staff during his chairmanship. The painting of Claude Chadburn which was to hang in the Brewery Chairman's Office for many years was painted at the Swan. Ivy recalled that the artist stayed at the hotel and was ferried to Papplewick Hall every day until the painting was completed. By the time Ivy left in 1970, to take up a new job in the canteen at Kings Mill Hospital, the more sedate world of banquets, dances and balls at the

Swan had been replaced by the disco.
In this same spirit of respect for employees past and present, the Brewery paid homage to its retired tenants by often giving them a special luncheon in the resplendent Brewery Cellars. One such event took place on August 25th 1981 when retired licensees with combined service totalling 200 years from such pubs as Mansfield's Kings Arms, Meden Vale's Three Lions and Mansfield's Reindeer enjoyed the hospitality laid on for them by the Chairman, Robin Chadburn. Although such occasions took place less than three decades ago, in retrospect it speaks volumes for an age, unlike today, when loyalty and long service were still respected in the corporate world.

FAKE FRENCHMEN & FEMALE FORKIES

Beaujolais Nouveau is always a wine surrounded by some odd traditions, as opposed to the more class-conscious vintages. Because everything depends not on how long this delicate liquid has been laid down, but on how soon it can be bottled, loaded, lifted and transported from the Beaujolais region in Burgundy in central France. Mansfield Brewery were keen contestants in the annual Beaujolais race. An example of their enthusiasm was outlined in the winter 1981-82 edition of the house magazine, *Marksman.* Derek Mapp, Mansfield Inns' Retail Director, appeared with the staff of the Racks and Casks Wine Bar all decked out as a superb 'ho-he-ho' fake Frenchman complete with striped jumper, beret and a suitable soup-strainer moustache as he delivered the first consignment of Beaujolais Nouveau for the lucky Mansfield clientele. In fact there was quite a bit of a French flavour around at the time, as *Marksman* reported that the licensees of Mansfield's Ravensdale Hotel, Bill and Barbara Toon, had been flown to France for a one-night stay after winning a draw run by a 'well known Brandy manufacturing organisation'. In Bill and Barbara's case the Frenchmen they met on their whirlwind trip were real – M. Jacques Hebard, Export Director, and M. Pierre Cordier, Managing Director of the famous Martell Brandy company.

MBC subsidiary T.W. Beach & Sons had been bought by the Brewery in September 1980, and had something equally as attractive as a young French wine in their busy depot. In what was generally considered a male-dominated occupation, Beach's possessed a trio of female fork lift truck drivers, Penny Nightingale, Nola Ellis and Beverley Teale. As *Marksman* put it; 'They have enjoyed their uplift… Penny enjoys her driving. All three handle their machines competently…'

THE MANDORA ROCKET

If you work for a company whose product range can often have the effect (if used in excess) of sending your consumers into orbit, then in retrospect it was hardly surprising that the shiny, tall, cylindrical Mandora tank and cooler (installed to provide greater CO^2 (carbon dioxide) capacity) should soon be generally referred to as 'the rocket' by staff – the comment at the time said it all; 'Improved conditions are favourable for lift-off – pity you can't see the crew…'

A CONTINENTAL TIPPLE…

There were lagers and there were lagers in the early 80s, but with a more sophisticated and better-travelled drinking public on the rise, the search had been on for a beer which actually possessed all the finer overtones of the European lagers. After extensive research MBC Chairman Robin Chadburn declared, on September 28th 1982 at the launch of the new Marksman lager, 'the significance of moving away from a 'home-grown' lager and the move to brewing Marksman to the continental method'.

The campaign to launch this highly successful brew was to cost a quarter of a million pounds and Richard Meadows, MBC's Brewing Production Manager, explained the subtle changes which had altered the lager giving it a more continental flavour and lighter colouring. At a second briefing, Richard Lewis, Group Managing Director, told the audience that 'the product's image should be correct especially as

the lager market was at present attracting mainly younger men who wanted their image to be 'right' with their mates.
In addition to a massive billboard advertising campaign, a whole range of Marksman products had been designed and manufactured which included sweat shirts, t-shirts, bar towels, aprons, ties, carrier bags, beanie hats, trays, ash trays, clocks, bingo boards, dominos, dart flights, golf tees, matches and luggage labels. The campaign obviously worked as within a year sales of Marksman were up by a staggering 42%, and the brew would go on to capture a good slice of the lager market in subsequent years.

WEDDINGS, COOKS & SNOOKER CHAMPS

We see that the attractive Mrs. Suzanne Drake, (21) started work as a cook in the Mandora canteen on the 18th of January 1982 after training at the highly-rated Saracen's Head at Southwell and earning her City & Guilds at West Notts. College in Mansfield. Did she go on to finer culinary achievements? We don't know… Another cook and West Notts. College City and Guilds success, Julie Vickers, of the Rufford Arms in Mansfield, was the runner-up in the CHAD newspaper's Hostess of The Year Competition, and won a £150 microwave oven (quite an innovation back then) donated by Baskills. Julie's menu for the competition was honeydew cup for the starter, main course of orange glazed lamb roast, duchesse potatoes, buttered carrots, peas and baby onions, and French apple flan with fresh cream as dessert – served, of course, with a bottle of Beaujolais Nouveau.
Stan Swindells, licensee of Mansfield's Reindeer Inn was the proud winner of the Conservative Club Snooker Handicap Challenge Cup, which was originally presented by the former MBC Chairman, Mr. Claude Chadburn back in 1957. Was there any inter-brewery competiton in this? Possibly – Stan defeated Bill Twells, licensee of the Tetley Miners Arms in South Normanton….

THE NEW BREWHOUSE

The 1980s was a time when the working population of Britain had to watch its back. By the second year of the Thatcher government, unemployment had increased by one million to reach a staggering 2,419,452, but although beer glasses may well have been the alternative receptacles of tears, the Mansfield brew was still as popular as ever. The company's annual report and accounts for March 1980 showed a new high in turnover at £25 million.

On May 14th 1980 the T.U.C declared a day of action during which workers throughout the U.K. would stay away from their work in protest at the increasing stranglehold the aggressive new government had on the unions. This act of protest was hardly the success the TUC had hoped for, however. Although a significant number of miners struck, many other industries had a near-normal working day. At Mansfield

The 150-foot high Brewhouse opens for business. Over 750 Mansfield Brewery employees from Mansfield, Hull and Evesham attended the successful Preview weekend on July 28th-29th.
Below: Chairman Robin Chadburn unveils the opening plaque with retired Directors Philip Baker and Edward Deakin.

Brewery, work went on as usual.

Throughout 1980 preparations had been well under way for the new brewhouse. The front page of the October 1982 issue of MBC's house magazine, *Marksman,* proudly displayed an architect's sketch of the proposed new brewhouse which, within a few short years, would become a familiar feature of the town's skyline.

The new building would be linked by an external staircase from the Great Central Road building and the fermenting rooms, with a new overhead conveyor from the malt bins. *Marksman* described how the new brewhouse would operate;

'Sweet wort in brewing is the extract obtained from the malt in the mash and is comprised mainly of sugar, proteins and amino acids. The present brewhouse has been producing worts for the past 30 years and it is hoped that the new one will produce them into the year 2000 and beyond. When considering the techniques to be applied in the new building the main consideration was that the Mansfield beers must still be of the very high quality consistently produced over the last century. Modern brewing techniques, some of which are used in the United States, were considered along with others but it was finally decided that the new brewhouse should combine modern but traditional plant with the latest computer controlled instrumentation.

This does not mean that the brewer and skilled operative will no longer be needed but that the tedious jobs will be controlled by the computer."

By August 1983 the stainless steel mash tuns and grist cases were being installed within the brewhouse's steel framework using a huge 110 tonne crane. The following month, the plastic-coated, re-lined cladding for the structure would be put in place. In October the final pieces of the equipment, including a mill for malt handling at 8½ tonnes per hour and manufactured by Vickers-Dawson of Crayford in Kent was swung up on the huge crane through the last open space on the brewhouse roof.

By September 1984 MBC Chairman Robin Chadburn unveiled a plaque commemorating the opening of the new, £4m brewhouse. Mansfield now had arguably the most modern brewery in Britain.

RESULTS: PROFITS UP DESPITE DIFFICULT CONDITIONS

The wonderful thing about the company's newsletter, the *Marksman*, was its informative mission to let employees know just how the company was doing. From labourers to supervisors, if you consulted the *Marksman* then you at least had an overview of what progress Mansfield Brewery plc was making. The August 1983 issue is a perfect example. It included an extract from the Chairman's Annual Statement;

"I am pleased to report a satisfactory year for your company in spite of continued difficult trading conditions. Profits before tax have improved by 11% over last year's 53 weeks, to £8.0 million, and trading profits increased by 13%. Turnover increased from £55.6m to £62m."

"The year saw a number of major property investments. The purchase of the Nag's Head, Belper, and the Seacroft Hotel, Leeds, has proved very worthwhile, exceeding all our expectations. A recent addition has been the Olde House, Chesterfield, which offers considerable potential for both pub and leisure developments. In addition we completed many alteration schemes reflecting our commitment to the continued improvement of our tied estates, the most notable being the Sir John Cockle, Mansfield, and the Chequers, South Emsall. During the year we sold the Station Hotel, Chesterfield and our squash club at Keyworth, Notts., as these properties no longer fitted into our developing leisure activities."

Success in The Press:

November 3rd 1983: Mansfield's *Chad* Celebrates

MANSFIELD BREWERY – ONE OF TOWN'S SUCCESS STORIES.

At the height of the Thatcher revolution big business stories were all the rage. The Miners' strike was yet to run its course and positive stories about private enterprise and entrepreneurship were filling Britain's newspapers. It was good news time for business, and the innovative Mansfield *Chad* , as ever, knew it had a gem on the doorstep with Mansfield Brewery. The paper devoted a massive four page spread to celebrate the company's long history. The first paragraphs of the text dealt with the earlier decades since the company's foundation, and local reporter Ronald Hodson brought things up to date with a glowing report of the way Mansfield's beer was made. As a snapshot of the Brewery at the height of its powers, this is hard to beat.

'The Mansfield pint is made from British barley that produces three qualities of malt required for the bitter, the mild and Marksman lager. Mashing, boiling, fermentation and conditioning are all essential processes that have been developed over the years to produce the special qualities required. Brewing, to the management and staff of Mansfield Brewery, has become more than just a profession but a way of life. It is notable for the number of long-service employees who work in modern, up-to-date conditions.'

'For 30 year-old Michael Evans in the Milling department, with 14 years service, and Dick Clarke and John Murphy, 15 and 19 years' service respectively and now operating in the Copperside section, each day can present its own problems in the running out of a good brew.

"We received an induction course when we started but the end product depends to a great extent on experience. Each day is not the same and experience counts a lot," Dick told *Chad.*

'In the matter of long service they are juniors compared with such as Fred Hensleigh, supervisor of the Bulk Beer Process, and with 48 years at Littleworth, or David Rhodes, painting supervisor, with 30, Walter Murray with 28 and Gordon Kendall, another supervisor, with 38 years. The Transport Department is well provided with vigorous veterans, such as Leslie Berry, 32 years service, Leslie Mayfield with 26, Albert Prest with 25 and John Eaton, 22, while the supervisor of the Bulk Beer Transport section, Eric Jackson, has been with Mansfield Brewery for 30 years.'

'At the other end of the scale one finds youngsters like 19 year old Adam Ferencz of Mansfield Woodhouse, with two months' service, and Mark Day (22), busily cleaning out the Yorkshire Squares after one week's work there.'

"One needs the ability to do manual work and to be adaptable. As fermenting is a living process there's always something different happening – a different problem every day," commented fermenting supervisor Roy Smith, who has been there nearly 25 years. The diverse activities at the Littleworth Brewery include a full maintenance section – fitters and electricians – and Estates Department under Arthur Tomlinson that has distinguished itself not least by its recent re-design and re-furbishing of a number of Mansfield pubs, and an office layout with open plan working space well abreast of contemporary office planning.'

'With around 250 employees on the payroll the Brewery has a large and dedicated staff, each skilled in his or her own particular application, though many have wide ranging skills covering most aspects of the brewing business.'

'The Group includes subsidiary Mandora (UK) Ltd., with an extensive soft drinks manufacturing capacity.'

'Group Chairman is Robin Chadburn, and Group Managing Director, John Hings.'

1983: CRASHING ADVERT FOR MARKSMAN LAGER

When part of the masonry from the top of Rotherham's multi-storey car park fell away and crashed onto the roof of a double decker bus below, it may have crushed the upper deck (thankfully no one was on board and driver Harry Hudson was unhurt) but it failed to obliterate the huge Marksman Lager advert on the side of the bus which, due to the drama of the situation, managed to be featured on TV screens to millions of viewers. Such free advertising was, and still is, priceless.

'AIR CONDITIONED' LOOS A THING OF THE PAST.

On Tuesday February 1st the New Inn on West Gate was packed with customers to celebrate the refurbishment of the pub. Gone were the old outside toilets, (natural air conditioning) and the new pub interior impressed everyone present. In toasting the success of the pub in the future, Group Chairman Mr. Robin Chadburn told the guests that the pub had been bought in 1897 for £4,168, whilst the 1983 renovation had cost £55,000.

SUMMER SUN SHINES AS THE BREWHOUSE OPENS – AND THE BALLOON GOES UP!

1984: More than 750 guests made the special visit to the Brewery during the preview weekend (28th-29th July) for the new, 150-foot tall Brewhouse. Lunches included salads and cold meats, complemented by Mandora soft drinks and cold beers, especially the specially brewed Number 3 Celebration Ale. Children enjoyed clowns, (a commodity the Chairman thought the Brewery already had enough of), and ice cream and played happily in a creche in the Brewery cellars.

Visitors came from all over, from Grimsby, Evesham and other locations, and for many this was the first time they had seen the Brewery's headquarters.

To complete the week of celebrations, shareholders, the press and friends of the Brewery from many branches of industry made their own tour of the plant marshalled by a team of thirty guides.

UP, UP…AND AWAY…

The only casualty of the brewhouse celebrations was the company's new promotional balloon. It had been purchased by Mansfield Inns for display at new pub openings and other events, and was making its first appearance at a group function – but it didn't last long. The 20 foot long balloon slipped its mooring rope and enjoyed a brief flight of freedom across Nottinghamshire and Lincolnshire before facing a shocking end on an electricity pylon….

GEORGE BUSH RETIRES…

George Bush, Mandora's personnel manager at Evesham retired after being presented with gifts from colleagues and staff to further his interest in photography – he had already been presented with a cheque and a camera the previous day by Mandora Managing Director Chris Rainsford.

1984: STRIKING MINERS BRING PLEASURE IN PLEASLEY

Bored strikers Mick Lapko, John Weaver, John Morton, Jack Davies, Derek Ward and Chris Skoponski weren't prepared to sit out the strike so decided that they'd do something about a rubble-strewn rubbish patch at the rear of their local, Pleasley's White Swan. With no previous gardening or landscape experience, the six pitmen transformed the blighted area into a tranquil, tiered garden, which became the joy of elderly people in Pleasley.

1987: GEORGE REMEMBERS BEING 'IN THE MONEY'.

George Powell, the Brewery's wages office supervisor, had been 'in the money' all his working life. Prior to life at the Brewery, where he joined in December 1961, he'd completed 13 years as wages clerk at Sherwood Colliery. During his National Service wages seemed to come into his life – he had to pay men off as a 'demob' clerk. George outlined some vivid memories of pay and conditions in the early days of his service with Mansfield Brewery. For instance, the highest paid foreman in the production section in 1962 was paid £25.10s.6d (£25.52) per week for 42 hours. Drivers received £13.7s.6d (£13.33p) for a basic 42 hour week. Maintenance fitters received just £18.11s for 45 hours, and annual holidays were two weeks rising to two weeks and three days after 15 years service. It was also standard practice then for lads to begin their employment in the bottling plant but they were automatically dismissed at 18 unless they showed aptitude to be employed elsewhere in the Brewery.

November 1987: **GHOSTLY STIRRINGS IN YE OLDE RAMME INN**

Renovations at Ye Olde Ramme Inn, Mansfield, could have disturbed a ghost – according to a Clipstone medium!

More than three years after landlady Annette Kolesniak was told of a secret tunnel, builders have discovered a hidden opening to a vault-like area.

"The medium said that the pub was haunted by monks who used the tunnel to get to and from the nearby St. Peter's Church," said Annette.

Local historian David Bradbury said that the pub was once used as a meeting place for a rebel Chartist movement and that the idea of a tunnel was feasible, however, it was more likely to be the entrance to the original cellar.

Work began on Mansfield's oldest pub to relocate the licensee's living quarters, making way for a separate games room next to the lounge bar. But as work progressed workmen unveiled a series of historical features pointing to the pub's age. Ancient wattle and daub walls and a free standing pillar in the entrance hall date back as far as the 1660s.

"We also discovered original beams when we removed a false ceiling in the lounge bar, and an old fireplace was revealed when we knocked down a wall," said Michael Kolesniak.

All of the findings have been preserved as features of the pub and enhanced by the new solid oak bars in both the lounge bar and games room.
The new games room houses a pool table and dart board and the pub has been redecorated. The ghost however can 'rest in peace' as the mystery entrance is being covered over once again.

SERGEANT TOM

George Powell recalled that there was a uniformed Sergeant-at-Arms named Tom Purdham who was on duty in the Brewery's reception on Littleworth Corner. Sergeant Purdham was a member of that elite corps, the Royal Corps of Commissionaires, and his other responsibilities included manning the switchboard "when each number had a different plug".

KING OF CLUBS:

George Powell was also instrumental in reviving the Brewery's sports and social club at New Houghton, and was a founder member of the Bowls Club. One of the more unusual club activities – considering the environment in which he worked – which George began was the Brewery's Home Made Wine Club, with up to ten employees bringing along their brews to taste at each session.

1988: CHARTING THE HISTORY OF THE BONNY BOAT

Landlord Chris Doyle of the popular Hull hostelry The Bonny Boat had been charting the history of the pub and appealed for local people to help. Originally the Bonny Boat was the meeting place for Merchant Navy sailors and the bargees who came to Hull down the river. The pub's original name had been the Bank Hotel, but when several banks opened nearby the name caused some confusion and was changed. Said Chris, who ran the pub with his wife Jean, 'The

Bonny Boat was an Eskimo kayak brought back by a 17th century sea Captain. The boat is now in the Trinity House Museum. I'm planning to get it photographed and display the picture alongside an old poem about the rescue of the Eskimo and his boat.'

The pub had just been refurbished in a distinctly nautical fashion. Lots of brass, and copper, plus a harpoon and even some old oars were on display.

THE FINAL COUNTDOWN

The 1990s

1991: THE POOL ROOM PHANTOM PULLS A FAST ONE.

There were spooky goings on at the Parliament Oak in Mansfield Woodhouse. The previous Christmas licensee David Tweddle had been awoken at 3.30 am by the pub's burglar alarm. He came downstairs to find … no burglar…yet all the lights were on in the pub 'Yet all the light switches were in the 'off' position!'
A regular had also once seen a figure of a woman walk straight through the wall of the gents. The most eerie yarn from the licensee was of the time he saw the pool table moving. 'I was expecting someone to be in the room but when I unbolted the door I saw the pool table moving across the floor. It took four of us to get it back the next day….' It gets spookier. Apparently, before it was a pub, the Parliament Oak was a cottage and when it first became a hostelry the Lady of The House hanged herself – with barbed wire. Surely, the customers can't have been *that* bad…Well, if you can't find spirits in a pub, where *can* you find 'em!

1992: LEARNING 'THE MANSFIELD WAY'.

New tenancies had meant a significant increase in workload for the personnel department. John Else, now Director of Personnel, dedicated a team to help integrate the new tenants into the brewery. Each new tenant was made aware of the wide range of training courses available, ranging from food hygiene to financial management, marketing and practical bar catering. Courses were also available which led to the qualification of the British Institute of Innkeeping. John Else said: 'We've spent a great deal of time interviewing both tenants and prospective tenants. Existing tenants have to be brought into the special Mansfield way of operating and those who have ambitions for a tenancy have to be carefully screened for the particular qualities we know we need...'

The Brewery Independent June 1993

Alongside the headline news that the Brewery's expansion programme had paid off handsomely, with pre-tax profits up 14% to £12.5 million, there was a more sombre note inside the newsletter from the new Chairman of Mansfield Brewery plc., Sir David White. It seems pointless to include the whole article here, considering the short span of independent life the Brewery had left at the time. Yet the tone amply demonstrates how much business and corporate affairs had altered since the political turnaround of 1979. The more benevolent, family atmosphere generated in earlier company newsletters seems to be waning.

RUSSIAN AROUND.....

Over in Hull at favourite sailors' and fisherman's pub, the Bonny Boat, multi-lingual landlord Wally Eckles was adding to his stock of languages with a study of Russian. Not bad for a bloke who had left school 40 years previously with no qualifications.
He'd already mastered phrases such as 'You're very welcome', and 'Thank You' for the visiting Russian sailors, who frequented the pub regularly. Wally also spoke Dutch and was fluent in Afrikaans, He was also very popular with visitors from abroad and out of town by giving away free maps of Hull's Old Town. Said Wally;
"British pubs have the reputation of being the most welcoming and friendly in the world – I'm just helping to make sure it stays that way. It's our job to make any weary traveller feel better – and greeting a foreigner in their native tongue is bound to make them feel welcome. Our Russian visitors really appreciate the efforts we make for them".

BREWERY BONFIRE PARTY A ROARING SUCCESS

1995's staff and families' Bonfire Party was anything but a damp squib and threw up some statistical fireworks which would ignite plenty of surly looks and land any of today's more frugal company accountants with a board room rocket. Lighting up the November sky, the 620 Bonfire attendees managed to consume:
176 gallons of beer, 960 bottles of pop, 650 hot pies, 40 lbs of gravy and an amazing 218lbs of mushy peas. There were rumours of strong winds blowing in Littleworth the following morning…

SAVING THE ROVER'S RETURN.

ITV's *Coronation Street* was as popular as it's ever been in 1995 when the writers had written a difficult time into the plot for Rovers Return landlady Bet Gilroy. Bet was in line to be evicted from the famous pub, so to the rescue came Mansfield Brewery plc, with the company mailing an offer to Bet of £80,000 plus a delivery to Lancashire via the company's 'Export Division' if necessary. Director of Sales Dennis Foster said "Bet is a national treasure and one of the best landladies in the country. The world should know of our concern for her and this wonderful neighbourhood pub. A great example of Serving Quality". We're not sure if Bet got the money or the delivery, but something ought to be done in the 21st century to get some of that Rovers quality back. Sadly, there's no Mansfield Brewery, as of old, to offer a helping hand….

THE LONELY WIVES CLUB…

By 1995 the Brewery had a maintenance staff of 17 dedicated men led by Mike Cleator, who had joined the company in the mid-70s. In addition to making sure the technical side of the brewery kept running smoothly, they were continually up-dating the systems. The two teams – electrical and mechanical, were led by Alan Singlehurst

and Ken Smith respectively. Mike Cleator recalled installing the PET plant in the early 1980s. That old dedication to making beer comes through again, something many ale widows of MBC's staff would recognise;
"The plant was originally created from second hand parts which had to be stripped down and re-built. It was a long job and sometimes our wives thought we were more attached to the brewery than to them!"
There's more than a grain of truth in that – but someone had to do it...

THE MANSFIELD MARTIANS HAVE LANDED....

In 1995 the BBC TV motoring show *Top Gear* had one of its most popular presenters. Jeremy Clarkson was at the top of his career. Acerbic, bordering on non-political correctness, opinionated and funny, Clarkson was the man of the moment, the kind of guy any 90s man would welcome sharing a pint or two with in his local, in fact the perfect man to appear in a Mansfield Brewery TV commercial. With some quirky metal Martians to share his pint, Jeremy's ads were soon the talk of every Mansfield pub.

OLDE WORLDE THEME AT BLACKSMITH'S

An Olde Worlde theme has been adopted at the Blacksmith's Arms at Preston, near Hull, after a £60,000 renovation.
The original lounge and back bar have been made into one large room which landlord Harry Moore considers to have doubled the pub's capacity. The walls are adorned with old farming prints, plates and brasses giving the pub a real country feel enhanced by original beams and fireplaces.
"We've heard nothing but good reports from all our regular customers and trade has picked up considerably at the weekends and mid-week," said Harry. The pub has been completely refurbished and redecorated inside and out, and the toilets have been moved inside.

A CAR AND A HALF – WITH EXTRAS

Everyone has seen a convertible car, but not quite like the one at the Charter Arms, Rotherham. For the new feature at the pub's recently renovated disco bar, Cadillacs, is a Cadillac car that has been cut in half. Imported from Virginia, it has been customised to house the disc jockey's console, which replaces the engine, and also several other 'added extras'. Dry ice billows from the radiator grill and the headlights have been linked to a new sound-to-light system.
Cadillacs has also been given a new lighting system with the latest mini lasers and neon lights, making it the most up-to-date disco bar.
The upstairs and outside of the Charters Arms has also been spruced up – the lounge and exterior have been redecorated and a tiled, pitched roof replaces the original one.

GRUESOME FIND AT PROVIDENCE

Some families have skeletons in the cupboard but licensees Terry and Val Gibbons were surprised to find they had one buried deep in their pub's foundations.
The Providence at Hull closed recently for extensive renovations and workmen uncovered their gruesome find during work on the pub's structure. Police were called to the St. Stephen's Square pub and the human remains were transferred to the Hull City mortuary. Further investigations revealed that the skeleton was over 100 years old and more than likely came from the burial grounds of a church which stood on the same spot before the Providence was built. The Providence re-opens after its major facelift has been completed on December 11th.

TOUGH TIMES AHEAD

'Britain's biggest brewers are seeing profits savaged by a price war that is allowing independent pub operators to get discounts of up to 50%. In recent weeks Bass, Whitbread and Allied Lyons have all

reported reduced profits. These figures make our results look even more impressive, but there's no room for complacency.
At the heart of the problem is an industry operating at 80% of capacity. To put that right would mean each of the six biggest brewers closing one brewery, or a weaker, smaller regional being forced out of the race. Our board is determined we will remain an integrated brewer - but that's going to be tough going. All the old established criteria will have to be revisited........we have also committed ourselves to becoming an Investor in People company. We have audited what needs to be done – and there's a lot!
We're introducing Profit Related Pay (PRP). In the future staff – from the board down – will get pay rises linked to company prosperity rather than inflation. The system, which doesn't yet include retail staff, is linked to our performance from May 1, 1993. The first payments will be made in October 1994, and as we're sharing the tax relief given to PRP schemes by the Inland Revenue, it should mean everyone getting a little bit more. But first we've got to make the profits...... we can do it, but it's going to be tough..'

No Names, No Pack Drill.
True Tales – But Mostly Anonymous

There are some ex-employees who have the odd story worth telling yet still feel the potential disciplinary ghost of their employment years looking over their shoulder. There are others who, although they enjoyed the social culture at the brewery, are loath to give the impression that they spent more time sampling the product than actually making it. Although the decades have passed, we have, by mutual agreement, ensured that these storytellers remain anonymous, but the stories are true, and in a book like this, too good to miss. Many other ex-employees may well recall the incidents in question; so, dear reader, if you're one of them, there's just one reason behind these revelations – sheer affection for a lost way of life .

How To Hide Your Supper

First, a story involving Head Brewer Michael Parsons;

"Parsons was famous for going out on the ale after work and I remember him getting a bloke into trouble one night. This poor employee was a bit hen-pecked, and he and Parsons had had a few and knew they'd be late home. So they stopped at the chippy and got some fish and chips. So this poor bloke gets home, and rather than annoy his wife, who had probably made his dinner for him anyway, he decided to hide his parcel of fish and chips in his raincoat pocket. He got a real ear-bashing from his missus on the doorstep, and daren't mention the hidden supper. He just took his mac off and hung it up in the wardrobe. The fish and chips were still there a few weeks later – nice and mouldy. All because of one of those late drinking sessions….."

Horses, Donkeys & Japs

Some employees remember early experiences of company cars.

"We used to get our cars filled up by Horace Redfern at his garage across the road from the Brewery. One of the lads, Terry Johnson, had a Datsun Bluebird but Horace wouldn't put any petrol in it because it was Japanese...Horace was a prisoner of war under the Japanese. He wasn't really a xenophobe – he just hated the buggers... but he never did fill that Datsun – he refused to do it."

"I remember being in Horace's one morning and a bloke came in from the Electricity Board, whose offices were just nearby. He wanted his car servicing, so he asked Horace 'Do you do any discounts for the Electricity Board?'

Horace replied 'Oh, aye, I do – same as you effin' do for me – *none!*'

'Ich bin ein Mansfielder' – Don't Mention The War...

Another story concerns the Brewery's hospitality to visiting dignitaries. There were occasions when mutual trading agreements might require a visit to Mansfield by other brewers.

"We had a party of visiting Germans from Bavaria who were coming to see the brewery one afternoon. We arranged to entertain them in the cellars. We decided it would be nice to make them feel at home, so we hired one of the local colliery brass bands to play. The band got into their stride and played well, and then topped their spot off by playing *The Dambuster's March.* Heaven knows what the Germans thought of this. Later on we had a quiet word with the band master.

'The *Dambusters?* What were you thinking of – wasn't that a bit insensitive?' The bandmaster just shrugged his shoulders;

'We've rehearsed the bugger so they had to have it!'.

Getting the Coal In....

Directors of the Brewery came and went, but those who retired could still have the staff at their beck and call. One example, as told by another employee, outlines what a different industrial world we live in today;

"Edward Deakin had been retired for about ten years when I got there, but I remember one winter's day it had been snowing quite heavily and we got a phone call from him. Michael Parsons took the message and he got hold of Malc Jenkes who was in the fermenting rooms – which, I might add, were once the company's stables – very hygienic! – and told Malc 'Go and get one of the mini-vans and get up to Edward Deakin's house – and you'd better shove a shovel in the back, because he wants you to fetch the coal in…' So Malc drives out to Deakin's place expecting to find a ton of coal on the pavement. But there wasn't. So he picks up his shovel and walks up to Deakin's front door and knocks on it. He comes to the door, opens it, and carrying a coal scuttle, says 'Ah, there you are. Here's the scuttle – the coal's round the back – go and fill this up!' So this bloke had knocked off work and driven there in a blizzard, armed with his shovel, whilst all the old man wanted was a bucket of coal for his fireside. Mind you, Deakin himself was a bit of a practical joker at times. There's one story of someone, during the war, when rationing was on, who had left a dozen eggs in an egg box in the office. Deakin hard-boiled them all and put them back in the box! It must have been quite a surprise when that person got the eggs home".

Hops, Health and Safety

"Deakin also had an unusual angle on health and safety in the old days. There was a bloke called Harvey Johnson once who was clearing the hops out of the old hop tanks behind the old brewhouse. This was the time before we had steel toe caps – he was wearing wellington boots and using a fork to shove the spent hops down through the hole into his wagon below. Well, he shoved a prong of the fork straight through his foot. So he's standing on the loading dock below the brewing office, he's taken his wellie off and starts looking at this blood-soaked foot and wondering what to do, when Deakin walks out the office door and shouts 'What've you done, Johnson?' to which Harvey replies 'I've stuck this fork through me foot!'

'Well, you silly bugger!' shouts Deakin, 'go and get it sorted out and *get them hops finished!"*

The Beers that Time Forgot

Gordon Kendall remembers some of the beers made by Mansfield in his early days. "There was a bitter, but mild was the big seller, and there was a nut-brown and a strong beer called Golden Drop – the last time they brought it out it was on sale at half-a-crown (12½p) a pint. But it just wouldn't sell. And we did pale ale and a stout in bottles…" Mike Cleator remembers a particularly 'special' brew:
"We did a commemorative beer once for Princess Anne…. trouble is we lost the recipe for that…"
Was there a disastrous beer in the Brewery's history? Everyone agrees- it was called 'RedEye'. Bob O'Meara explains:
"RedEye was a marketing disaster. This is where we got lulled into thinking that because there was a taste for creamy Irish beers – Caffreys led the field at the time – that we had to have one. The name for a start was wrong. It sounded like a bad air trip or an overnight train ride – or even a disease. We spent absolutely thousands of pounds trying to shift RedEye…"
Mike: "The beer itself was alright. It was like a lager but fermented with an ale yeast – but we just got the marketing all wrong. We put the money in the wrong place – and we didn't tell anybody what the beer was or what it meant. We were charging the punters more money for it but we didn't tell them why. And you couldn't dispense it at speed – we didn't put enough money by for special pumps to serve it. It just didn't work…"
Bob: "Well, every brewery is allowed to make one mistake. RedEye came and went in less than a year – but even that doesn't compare to me losing the brewery that half million quid in Russia! Looking back, it's a good job I'd already made five million before hand – and I remember when I got to Russia the first time, they'd just had a revolution and a

day later the Russians invaded Chechnya – so I'm probably a bit of a historical common denominator…"

With All Due Respect…

John Else might have been Head of Personnel but based on the evidence it didn't always cut all that much ice with his peers.

"I used to attend these technical seminars at Burton, but I didn't understand a word of what was being said. My job was to drive Brewing staff there and that was probably due to the fact that I had a company car. Well, there were four of us going down the A38 one night, and we had Radio 4 on. There used to be a schools quiz then – *Top of The Form,* where bright kids in schools answered questions in teams – a bit like a junior *University Challenge.* My three passengers were feeling a bit confident because they'd actually got two questions right. Then the quiz master asked;

'*How do you spell the word 'Personnel'?'* and spontaneously, in unison, these three went "W- A- N-…" well, there was a 'K' in the middle and it ended in 'R'. Talk about 'respect'? What a joke!"

Don't Bark at The Bishop…

Respect in hallowed company reminds Bob of one of Mansfield Brewery's female employees who worked in the computer and IT department. Her name is Marie, and by all accounts this fine woman had more than one fruity turn of phrase.

"I'd noticed that she'd got this little puppy, and she seemed a bit concerned. She told me 'He's got some sort of blisters all over his tongue and the vet's given me this ointment, but I can't get the little bugger to keep the ointment on his tongue…' I met her again a couple of weeks later and asked her how the pooch was.

'Oh, his tongue's cured now….' I asked her how she'd managed to make the dog keep the ointment on his tongue.

"He didn't – I just smeared it all over his bollocks – I knew he'd be

licking them eventually – that cured him!"
Some time later, when the dog was a little older, she used to take him everywhere. After the 'Save Mansfield Brewery' march we all attended a function in Mansfield which ended up at a session in the Civic Centre where the Bishop of Sherwood was one of the presenters. About three rows from the back, there was Marie holding her little dog. Now the dog wasn't all that interested in the Bishop's talk so he decided he'd contribute to the occasion by barking. He held the centre of attention for a while until we all heard Marie say 'Shut the f--- up! That's the F----ng Bishop of f----ng Sherwood speaking!' "

❖ ❖ ❖ ❖

The Hangover – Last Orders!

So, ladies and gentlemen, it's 'time, gentlemen, please!'. We hope you've enjoyed this somewhat unusual and, in places, irreverent history of one of Mansfield's most loved businesses.

The 1994 Brewer's Journal Challenge Cup for the Best Brewed Beer in Cask Presented to Mansfield Brewery for Riding Bitter and Associated Medals for other brewing achievements.

When we stand at a bar and look at the badges on beer pumps we're not always necessarily thinking beyond the taste of our chosen tipple. Yet behind every badge of every brewery there's a story. One thing's for certain – Mansfield's story is unique in so many ways, from the child-like innocence of poor Frank Baily Thomas through to the hilarity enjoyed by the brewers at Littleworth, this has been a story where the old adage 'you couldn't make it up' has a genuine ring of truth.

Needless to say, with literally thousands of employees passing through Littleworth and hundreds of landlords, landladies and bar staff frequenting hundreds of Mansfield hostelries down the decades, we know that we've only scratched the surface with this book. There's a lot of stories as yet untold. So, here's the deal; if you've enjoyed this book and, as an ex-brewery employee you're feeling a little disappointed that your name doesn't feature, then let us know. If this volume has any degree of success, then we feel there's always room for volume 2. Landladies, landlords, draymen, barmaids, in fact anyone whose life has been graced by Mansfield Brewery – if you've a story to tell, get in touch. In a century we'll all be history in any case – but we'd

like to leave something for beer drinkers of the future to ponder over. We're losing all our family breweries, but there is hope – today's micro breweries could well become the family breweries of the future.
So, to all of you who have contributed to this book, and for those still to make themselves heard, we raise a glass of the famous brew in your honour.
And always remember – not much matches Mansfield!

❖ ❖ ❖ ❖

To all Mansfield Brewery Group Employees and Licensees

December 1979

FOR YOUR RETENTION

MANSFIELD BREWERY COMPANY LIMITED

Company Objectives

1. To plan for the long term survival and development of the Company as an independent public company with a distinctive management style, where individuals matter and can enjoy working.

2. To increase profitability consistent with the achievement of objective 1. and to show everyone how their own efforts contribute to the Company's success.

3. To give the utmost consideration to the needs and interests of our licensees and their customers; to maintain the high quality of the Company's products and outlets, and to retain our leading position as the most highly respected Brewer within our trading area.

4. To ensure job satisfaction for all employees and licensees through enlightened management and to encourage their contribution to and benefit from the profitable growth of the Company.

5. To keep employees and licensees informed of policies, progress and problems and to invite their comments and criticisms.

6. To improve working conditions wherever possible and to take all appropriate steps to ensure the health and safety of all employees and licensees. To promote continuity of employment by careful forward planning and by the early recognition of the effects of change.

7. To give Management appropriate freedom of action so that they can personally influence profits. To train and develop employees and licensees and to promote from within when appropriate.

8. To be flexible and not depend too much on any one product, customer or market. To maintain an efficient research and development policy so that opportunities can be quickly recognised and speedily exploited.

9. To work to benefit the local community and to improve the quality of the environment.

W&DB News Item
BREWERY REVIEW
Release date: 25 April 2001 @ 09:00

Closure of Mansfield Brewery

The brewery operation will close by the end of 2001. The Group will continue to supply Mansfield beers, which have a strong regional franchise. The brewing of Mansfield beers will transfer to Park Brewery, Wolverhampton under the guidance of the current Mansfield Head Brewer, Robert Burton. He will be responsible for maintaining the excellent quality of the Mansfield ales and ensuring a smooth changeover. There are no changes planned for the Mansfield beer range.

The 2,000 people employed in our pub business across the East Midlands and Yorkshire area are not affected by these changes.

The use of the Mansfield name and pub signage will be unchanged. The Group will continue to invest in its pubs and free trade business in the Mansfield trading area, which is a key part of the future plans announced today.

INDEX

Figures in italics indicate captions. 'MBC' indicated Mansfield Brewery Company.